D1090469

/ 777

MEASURES OF ENVIRONMENTAL PERFORMANCE AND ECOSYSTEM CONDITION

Peter C. Schulze, *Editor*

NATIONAL ACADEMY OF ENGINEERING

NATIONAL ACADEMY PRESS
Washington, D.C.   1999

NATIONAL ACADEMY PRESS • 2101 Constitution Avenue, N.W. • Washington, DC 20418

The National Academy of Engineering was established in 1964, under the charter of the National Academy of Sciences, as a parallel organization of outstanding engineers. It is autonomous in its administration and in the selection of its members, sharing with the National Academy of Sciences the responsibility for advising the federal government. The National Academy of Engineering also sponsors engineering programs aimed at meeting national needs, encourages education and research, and recognizes the superior achievements of engineers. Dr. Wm. A. Wulf is president of the National Academy of Engineering.

This volume has been reviewed by a group other than the authors according to procedures approved by a National Academy of Engineering report review process. The interpretations and conclusions expressed in the papers are those of the authors and are not presented as the views of the council, officers, or staff of the National Academy of Engineering.

Funding for the activity that led to this publication was provided by the Andrew W. Mellon Foundation and the National Academy of Engineering Technology Agenda Program.

Library of Congress Cataloging-in-Publication Data

Measures of environmental performance and ecosystem condition /
edited by Peter C. Schulze / National Academy of Engineering.
   p.   cm
   Includes bibliographical references and index.
   ISBN 0-309-05441-9
   1. Environmental indicators. 2. Environmental
management—Evaluation. 3. Environmental monitoring—Methodology.
4. Nature—Influence of human beings on—Evaluation. I. Schulze,
Peter C. II. National Academy of Engineering.
   GE140 .M43 1998
   363.7'063—dc21   98-43781

Cover: George Inness. *The Lackawanna Valley* (detail), 1856. Oil on canvas. National Gallery of Art. Gift of Mrs. Huttleston Rogers.

Printed in the United States of America

Preface

Like all other species, humans impact their environment. The scale of human impacts has grown as a result of population growth and increased consumption of goods and services. At the same time, our understanding of the environmental consequences of human activities has improved.

Decades ago, attention focused mainly on clear-cut, obvious environmental insults: deadly chemical fogs, burning rivers, and eutrophic lakes. Today, scientists and the public are paying more attention to less apparent impacts such as stratospheric ozone depletion, greenhouse gas emissions, bioaccumulation of toxic chemicals, and the disappearance of unfamiliar or even unknown species.

This broadening appreciation for less obvious but still significant environmental impacts has elevated the importance of methods for detecting and measuring substances known to affect the health of the environment. Currently, dozens of measurement techniques are in relatively early stages of development or adoption. Some are intended to help study the condition of an ecosystem; others are designed for comparing the impact of alternative human activities.

The two categories of metrics have been developed by two cadres of professionals: those focused on assessing the condition of ecosystems and those interested in assessing environmental impacts associated with particular activities or products. Although these two groups play complementary, closely related roles, they have traditionally had little interaction.

The papers in this volume are the product of a 1994 National Academy of Engineering (NAE) workshop. The workshop was intended to promote interaction, coordination, and crossfertilization between those who assess and manage the condition of ecosystems and those who assess and manage the environmental performance of institutions. The papers were contributed by engineers, ecologists, managers, and academics. Each discusses a particular approach either for assessing the condition of ecosystems or for assessing environmental performance. This volume does not attempt to present a comprehensive review of the multitude of assessment techniques presently under devleopment. Rather, it provides an introduction to these two closely related fields by highlighting key features of some of the more prominent approaches.

The idea for the workshop grew out of discussions among NAE member Robert A. Frosch, who chaired the workshop, former NAE Program Office Director Bruce Guile, former NAE Fellow Peter Schulze, and Deanna Richards, who directs the NAE program on Technology and Environment (T&E). This volume and the workshop are components of NAE's ongoing initiative to explore issues of technology and the environment. We are indebted to the authors for their excellent contributions and to an editorial team composed of Peter Schulze, Greg Pearson, Penny Gibbs, Long Nguyen, and Jessica Blake. Peter Schulze was also assisted at Austin College by the careful work of Stephanie Hinds, Lanell Tweddle, and Amberly Zijewski.

Finally, I thank the Andrew W. Mellon Foundation for its generous support of the NAE's Technology and Environment program. This funding was critical to the success of the workshop and the completion of this report.

Wm. A. Wulf
President
National Academy of Engineering

Contents

v

INFORMATION FOR MANAGERS

BIOASSAYS

Measures of Environmental Performance and
Ecosystem Condition. 1999. Pp. 1–12.
Washington, DC: National Academy Press.

Overview:
Measures of Environmental Performance
and Ecosystem Condition

PETER C. SCHULZE AND ROBERT A. FROSCH

*". . . everything is an indicator of something but nothing is an indicator of
everything."*

(Cairns et al., 1993, p. 6)

No metrics were necessary to recognize a problem when the Cuyahoga River
caught fire in Cleveland. Likewise, any casual observer would realize that something is wrong with the Aral Sea, where commercial fishing vessels lie stranded
in their ports dozens of kilometers inland. In both cases, shortsighted human
behavior led to dire, readily detectable environmental consequences.

The problem with relying on such blatant evidence is that environmental
deterioration becomes critical before a response is even contemplated, let alone
implemented. Rather than wait for disasters to happen, one would prefer to avoid
problems in the first place. This, however, requires the ability to predict the
environmental consequences of human activities, avoid activities with unacceptable impacts, and document the environmental consequences of other activities.
Any sophisticated attempt to predict and detect the environmental impact of human activities requires appropriate measurement methods.

Since the modern environmental movement picked up steam in the 1960s,
we as a society have effectively addressed some of the most obvious or straightforward environmental effects of human activities. Gasoline sold in the United
States no longer contains lead, hence, there is less lead in the air. One rarely sees
a belching smokestack. The Cuyahoga River, much cleaner now, is not likely to
catch fire again.

The human impacts on the environment that receive the most attention today are often either obscure or complex. Human impacts on stratospheric ozone and greenhouse gases provide examples. Unlike a belching smokestack, the effect of chlorofluorocarbons on stratospheric ozone is apparent only to those with the expertise to understand the critical atmospheric chemistry. Meanwhile, a plethora of human activities contribute greenhouse gases to the atmosphere and thereby have the potential to alter climates. As a result of such obscure and complex relationships, it is difficult to predict the future environmental consequences of human activities.

As more obscure and complex relationships between humans and the environment are recognized, those charged with reducing and monitoring human impacts face greater challenges. They must shift their attention from obvious and relatively simple impacts (e.g., lead in gasoline or oil floating on a river) to more complicated processes (e.g., the effects of manufactured chemical compounds on animal development and human health [Sharpe, 1995]). Thus, they need to improve their ability to predict the ecosystem consequences of changes in human activities. As usual, better predictive abilities will require better measurement abilities.

One can envision a continuum of progress, from a past when there was little concern for the obvious environmental impacts of human activities, to the present when some obvious effects have been reduced but new understanding has led to new concerns, and, finally, to a future when comprehensive understanding of the environmental implications of human activities makes it possible to eliminate any activities that have unacceptable environmental consequences. The latter is probably too much to expect, given the diversity of ecological impacts, the potential for interaction between these impacts, and the complexity of the ecosystems involved, but it is a worthy target.

Over the past few decades, two new groups of professionals have emerged. One is responsible for managing and reducing the environmental impacts of human activities. The other is responsible for assessing and monitoring the conditions of ecosystems that are affected by human activities.

Those charged with managing and reducing environmental impacts have developed a suite of metrics for gauging environmental performance. Those who assess the status of affected environments have developed metrics for determining ecosystem conditions. However, surprisingly little interaction appears to take place between those working to improve environmental performance and those trying to monitor the condition of ecosystems. With so little interaction, those who work to improve environmental performance are rarely able to assess the marginal environmental effects of their improvements. Meanwhile, those who measure the condition of impacted ecosystems often lack information on the particular human activities that are responsible for changes in ecosystem conditions. As attention turns to more obscure and complex environmental impacts, the lack of communication between these two groups could substantially impede progress. How can a plant manager know which of two alternatives in a production process would be least harmful to the environment? How can a designer know whether a compound made of material x would be better or worse for the environment than a compound made of material y? In some cases, the answers to these questions are straightforward, but in many cases they are not.

Ideally, a change in environmental performance could be measured in units of ecosystem condition. For example, computer design might be assessed a priori in terms of its expected per-unit impact on, for example, the water quality in streams that receive effluent from the computer factory. Recent efforts have attempted to extend life-cycle assessment procedures toward this goal (Steen and Ryding, 1992), but those efforts have been criticized because of the difficulty of ranking the environmental significance of different types of impacts. In other words, there is inadequate information on how different impacts affect ecosystems. As a result, critics are unconvinced that improvements in environmental performance measures will correlate with improvements in ecosystem conditions (Field and Ehrenfeld, this volume). Absent any correlation, efforts to improve environmental performance could be ineffective or even possibly counterproductive.

In other cases, there may be insufficient information to confirm that adherence to environmental performance standards does safeguard impacted ecosystems. For example, federal regulations require the use of bioassays to measure the toxicity of effluents released directly into waterways (Goulden, this volume). The presumption is that if the effluent bioassay results meet the environmental performance standard, then the ecosystem will not be harmed. Goulden argues that this is not a safe assumption. This appears to be another case of insufficient collaboration and cooperation between managers of environmental performance and managers of ecosystems. Hart (1994, p. 111) notes that the Intergovernmental Task Force on Monitoring Water Quality determined that ". . . for every dollar invested in programs and infrastructure designed to reduce water pollution, less than two-tenths of one cent (or 0.2 percent) was spent to monitor the effectiveness of such abatement programs!" Field and Ehrenfeld and Goulden essentially argue that one can not assume that an improvement in environmental performance, as measured by life-cycle assessment or effluent bioassays, leads to even an incremental improvement in ecosystem condition. Clearly, there is ample room for better coordination and collaboration between students of environmental performance and students of ecosystem condition.

Although it is probably too much to expect a comprehensive ability to predict how each particular human activity affects ecosystems, existing evidence suggests that collaboration between ecologists and engineers can be a powerful means of simultaneously achieving engineering objectives and environmental goals. Shen (1996) and Lindstedt-Siva et al. (1996) describe relevant examples from the fields of oil exploration and water management. In both cases, explicit environmental objectives served as engineering design constraints. Environmental scientists identified precise design criteria that engineers then used. The projects described by these authors have not yet been completed, so convincing evidence of success must await studies of the actual impacts of those designs. Nevertheless, the designs appear to satisfy the particular ecological constraints that served as design criteria.

Strang and Sage (this volume) describe a similar collaboration, but one with a 30-year track record. Since 1965, Eastman Chemical Company (and its predecessors) has worked closely with aquatic scientists from the Academy of Natural Sciences in Philadelphia to improve the environmental performance of the Tennessee Eastman Division, a facility that releases effluents to the South Fork Holston River. Staff members of the Academy of Natural Sciences periodically assess the condition of the river. Eastman personnel then use the assessment results to help determine what specific objectives are appropriate for efforts to improve environmental performance. As Eastman modifies facilities and practices to improve performance, subsequent studies by the Academy of Natural Sciences document resulting changes in the river's condition. This procedure, which depends on a strong collaboration between corporate managers and aquatic ecologists, has led to a long record of continual improvement in both environmental performance and ecosystem condition. Eastman used what it learned from the Tennessee Eastman Division to design facilities on the White River in Arkansas. Studies of that river by the Academy of Natural Sciences have shown no differences between the ecosystem conditions upstream and downstream from the Eastman plant.

The close relationship between the properties of liquid effluents and the condition of receiving waters undoubtedly facilitated the success of the collaboration between Eastman Chemical Company and the Academy of Natural Sciences. Many other situations involve a less direct connection between environmental performance and ecosystem condition. Nevertheless, Strang and Sage's account confirms the potential for progress in environmental performance and ecosystem condition when corporate managers and environmental scientists collaborate closely.

This volume is intended to facilitate that progress by reporting on a variety of important metrics that have been developed to assess environmental performance and the condition of ecosystems. Our hope is that the discussion of these various indicators in one volume will foster not only further refinement of the particular measurement techniques but also more communication between users of the two sets of measures so that examples such as the one described by Strang and Sage will accumulate rapidly. The remainder of this chapter provides a brief summary of some of the important metrics that are used to assess environmental performance and ecosystem condition.

EXAMPLES OF ENVIRONMENTAL PERFORMANCE METRICS

Measurement methods are being developed for a variety of purposes, from tracking the impact of a soda can to gauging the performance of national economies.

Life-Cycle Assessment

Life-cycle assessments attempt to summarize the environmental impacts of a product through its entire life-cycle, from the extraction of the raw materials through manufacturing, use, and disposal. Such information can serve as the basis of a search for design alternatives to reduce the environmental impact of a particular product or to compare several different technologies designed to serve the same function.

Although simple in principle, life-cycle analyses can be difficult to complete in practice because they require large amounts of information and frequently involve assessments of the relative importance of qualitatively different types of environmental impacts (Field and Ehrenfeld; Hocking, this volume). Comparing two alternatives that have qualitatively different environmental impacts is one situation in which collaboration between engineers and environmental scientists could be most useful. Engineers have the expertise to develop design options. Environmental scientists will have more information (although often not enough) about the environmental consequences of releasing different wastes. Even if it will never be possible to predict precisely how different design options will affect the environment, it seems self-evident that life-cycle analysts would benefit from the insights of ecosystem experts and that ecosystem experts would profit from an understanding of design options and production processes. If nothing else, such information could help set priorities for ecosystem research.

Field and Ehrenfeld elaborate on the difficulty of putting life-cycle assessment into practice. They explain the serious limitations that arise due to the inability to rank the importance of qualitatively different environmental impacts associated with different technologies or design alternatives. These limitations notwithstanding, the authors emphasize that life-cycle assessments help to illuminate the differences in the environmental properties of various technologies.

Hocking argues that the problem of ranking qualitatively different types of environmental impacts can be solved by using energy requirements as a basis for comparing different technologies or alternative designs. He notes that differences in the emissions characteristics of two technologies can often be overcome by the expenditure of energy to reduce the emissions of the poorer performing technology. He illustrates the insights that can result from an energy-based life-cycle analysis by comparing the energy requirements of ceramic, plastic, and paper cups.

Allenby and Graedel (this volume) focus on the extensive data requirements of conventional life-cycle analyses. They argue that if data requirements are too great, life-cycle analyses simply will not be performed. They choose instead a qualitative checklist approach and show how it can be used to guide the site selection and design of corporate facilities.

Todd (this volume) notes that Allenby and Graedel's decision-support matrix is the type of tool that many managers lack. Managers may wish to identify means of pollution prevention, for example, but not have the information they need to identify opportunities and make good decisions. Todd explains that the information needed to make good environmental decisions may be either lacking entirely or unreliable. She proposes a set of guidelines for developing an effective information and measurement system to support environmental decision making.

Plant-Based or Organization-Based Measurements

Raw Flux Measurements

Perhaps the most important raw flux measurement system in operation is the Toxic Release Inventory (TRI). TRI was mandated by the 1986 Emergency Planning and Community Right-to-Know Act. The law requires companies to publicly report releases of each of more than 300 chemicals. Many people believe that the TRI has been remarkably effective in reducing environmental impacts. When companies and people living near them have learned of the quantities of emissions from plants, those companies have often chosen to voluntarily reduce their emissions. The beauty of TRI is that no one is required to act in response to the TRI data, but many have done so anyway. Critics have emphasized that the TRI does not distinguish chemicals on the basis of their relative toxicity and have questioned the accuracy of emissions reports.

3M's Waste Ratio

3M calculates a simple "waste ratio" to assess the environmental performance of their operations (Zosel, this volume):

$$(waste)/(waste + products + by products).$$

The waste ratio uses mass as a common currency that can be followed over time. The advantages of the waste ratio include its simplicity and its limited data requirements. Using mass balances, the waste ratio can be calculated from information on the mass of products, by-products, and wastes. Most of these data are already collected for other purposes, or can be calculated from existing data and process engineering relationships.

The major disadvantage of the waste ratio is that it characterizes waste purely on the basis of mass. A kilogram of nontoxic waste has the same effect on the waste ratio as a kilogram of highly toxic waste. Nevertheless, initial results suggest that the waste ratio can be a valuable tool for improving environmental performance. From the time 3M Corp. introduced the ratio in 1990 through 1995, it reduced wastes 32.5 percent worldwide.

Effluent Bioassays

Effluent bioassays measure the survival, growth, or reproduction of aquatic organisms exposed directly to effluents (Goulden). Effluent bioassays are frequently required by government regulation. Their advantage is that they directly measure the response of organisms to whatever mixture of materials is being released to the aquatic environment. Strictly speaking, these bioassays measure environmental performance, but they have properties more characteristic of measures of environmental condition. For example, like most biologically based measures of ecosystem condition, the test organisms respond to the complete suite of effluent components. Thus, with the use of effluent bioassays, humans need not make assumptions about the relative toxicity of different effluent components. However, it can be difficult to determine why the organisms respond as they do or even to what they are responding. Finally, as Goulden argues, the widespread use of effluent bioassays risks a sort of complacency if bioassay results are extrapolated carelessly. As Goulden explains, an effluent could damage an ecosystem even if it does not appear to be toxic in conventional effluent bioassays.

National or Regional Measurements

Natural Resource Accounting

From an environmental perspective, the widely used measure of national economic health, gross national product (GNP), has two important shortcomings: It does not include depreciation charges for depletion of the natural resources that form the basis of production, and it does not incorporate the costs of environmental externalities (environmental consequences of market transactions that are not reflected in market prices). Repetto et al. (this volume) explain that "a country could exhaust its mineral resources, cut down its forests, erode its soils, pollute its aquifers, and hunt its wildlife and fisheries to extinction, but measured income would not be affected as the assets disappeared."

Natural resource accounting methods attempt to overcome the environmental shortcomings of traditional accounting calculations (Daly and Cobb, 1989; Solorzano et al., 1991). These methods ascribe prices to various forms of environmental degradation and then add those values into conventional accounts, thereby providing more comprehensive accounts of the consequences of economic activities. Repetto et al. use a case study of Indonesia to illustrate natural resource accounting. They show that although Indonesia's GNP increased by an average of 7.1 percent per year from 1971 to 1984, the increase drops to an average of 4.0 percent per year when a few major forms of environmental depletion are factored in.

Repetto et al. further describe the insights from natural resource accounting that accrue when the method is applied to particular components of a nation's economy. They show that conventional accounting calculates the net economic value of a Pennsylvania corn and soybean farm as $75 per acre per year, but that the value drops to only $2 per acre per year when calculated by natural resource accounting methods. Finally, Repetto et al. demonstrate that estimates of industrial productivity growth can be spuriously low if they do not account for emissions reductions. Using natural resource accounting methods, they estimate productivity growth in the electric power industry to be two to three times as high as conventional estimates of productivity growth during the same period.

National Materials Use and Waste Production

Ayres and Ayres (this volume) and Wernick and Ausubel (this volume) focus on aggregate waste generation by the U.S. economy. Ayres and Ayres make detailed calculations to estimate this aggregate waste production. They use a mass balance approach that combines information on the import and extraction of materials in the United States with information on production of goods. They calculate waste generation by subtracting the quantity of materials in goods produced from the materials imported and extracted domestically. This is similar to 3M's system of calculating the mass of products, by-products, and wastes produced in its plants (Zosel), but it is applied to the nation as a whole. Such information can be instructive for a variety of reasons. First, measurements of this type provide a comprehensive baseline by which to measure future progress in waste reduction. Second, they tally the various types of waste production in the various industries. Third, these measurements may help to identify previously overlooked opportunities to reduce waste production because although data are regularly collected on materials extraction and production of products, rates of waste production often become apparent only through deliberate, detailed calculation.

Wernick and Ausubel suggest a variety of metrics that can be used to track changes in national environmental performance. Their metrics measure a variety of features of national materials use and waste production, such as aggregate consumption of materials per capita, ratios of uses of various fuels, consumption of materials per unit of economic production, growth-versus-harvesting ratios for natural resources, and inputs of agricultural chemicals per unit of agricultural production. Cox and Offutt (this volume) describe recent efforts to establish similar metrics for farming and ranching. Such measures should prove useful for tracking national and sectoral progress in waste reduction and identifying opportunities for improvements within various sectors of the economy.

Global Estimates of Environmental Performance: Appropriation of Primary Production

An interesting aggregate measure of global human impacts is Vitousek et al.'s (1986) estimate of the proportion of net primary production[1] "appropriated" by humans. This is one of the few performance measures that not only attempts to quantify total human impacts, but does so in terms that are related directly to the environment's potential for support. Vitousek et al. (1986) estimate that humans appropriate 25 percent of global net primary production and almost 40 percent of terrestrial net primary production. By appropriation, they mean the sum of direct use of plants for food, fuel, fiber, and timber and the reduction in primary production that would otherwise occur through alteration of ecosystems by deforestation, desertification, paving, or other types of conversion to a less productive condition. They conclude that ". . . with *current* patterns of exploitation, distribution, and consumption, a substantially larger human population—half again its present size or more—could not be supported without co-opting well over half of terrestrial NPP [net primary production]" (p. 373). Regardless of whether this estimate is accurate, human use of resources and the impact of such usage on the life of the planet is clearly massive and growing.

EXAMPLES OF ECOSYSTEM CONDITION METRICS

Like performance measures, a diverse collection of methods is now used to assess the condition of environments or their components. These range from simple physical and chemical measurements to measures of the condition of individual organisms and methods for assessing the condition of entire ecological communities. Some conventional ecological measures, such as the biomass or size structure of a population, are useful as measures of ecosystem condition. Others, such as the Index of Biotic Integrity (IBI) (Karr, 1981), were developed for the express purpose of measuring ecosystem condition.

An extensive literature describes the relative merits of various measures for various applications. We do not attempt to review that literature here, but merely describe briefly some important methods. Introductions to the literature are provided by Schindler (1987, 1990) and Cairns et al. (1993).

Physical and Chemical Measures

Environmental conditions have traditionally been assessed with physical or chemical measurements, such as the pH of water or the temperature of air. These measures can be very informative and will surely remain important. However, it is often difficult to predict the ecosystem consequences of a change in physical or chemical conditions. In addition, conventional physical or chemical measures do not always detect important changes (Karr, 1991; Yoder and Rankin, this volume). As a result of these limitations, there is a trend toward reliance on a more balanced combination of physical and chemical measures plus direct biological criteria for assessing ecosystem condition.

Measurements of the Condition of Individual Organisms and Populations

Physiological, Histological, and Demographic Measures

There are a variety of methods for assessing the condition of individual organisms that can provide evidence of environmental degradation. These include measurements of body burdens of various compounds (e.g., polychlorinated biphenyls, mercury), the prevalence of cancers or deformities, and the concentrations of enzymes that are synthesized in response to environmental contaminants. Other types of environmental impacts can be detected from studies of the size or structure of populations. These latter measures can be particularly useful for monitoring the status of harvested populations such as fish or trees.

Ambient Bioassays

Like the closely related effluent bioassays described above, ambient bioassays expose test organisms to a stimulus. In effluent bioassays, the stimulus is an effluent. In ambient bioassays, the stimulus is usually water from a polluted or potentially polluted source, such as a river. Like effluent bioassays, ambient bioassays measure the survival, growth, or reproduction of test organisms. Whereas effluent bioassays are used to assess the toxicity of particular effluents, ambient bioassays assess the cumulative toxicity of point and nonpoint sources of pollutants after their dilution by, for example, a body of water. Stewart (this volume) describes the insights yielded by ambient bioassays and the considerations that are necessary when evaluating ambient bioassay data.

Measurement of the Condition of Entire Ecological Communities

Particular species have long been used as "indicators" of ecosystem condition. Indicator species are organisms whose sensitivity to pollution makes them useful as a tool for detecting polluted sites. The concept is useful, but reliance on a particular species makes for a crude measure; the indicator species is either present or absent. Absence is not necessarily a result of local conditions; the species may never have had the opportunity to colonize the site. In addition, particular indicator species may help detect particular environmental impacts, but they are not likely to be suitable for detecting a wide range of impacts.

Information on the presence or absence of indicator species has frequently been supplemented with various basic ecological measurements such as species richness (e.g., the number of fish species in a particular section of a river), the abundance of various organisms, or more formal ecological measures of species diversity. Most of these measures assume that analysts have good information on the characteristics of relatively undisturbed reference sites in the region of interest. Strang and Sage demonstrate how such measurements have helped to document improvements in the conditions of the rivers they studied.

Index of Biotic Integrity

Recently, the concepts of indicator species and species diversity have been elaborated with the development of so-called multimetric biotic indices. These measures assess the overall condition of an ecosystem through simultaneous use of a variety of metrics. One such index is Karr's (1981) IBI. Karr (1992) argues that the multivariate nature of natural systems dictates that effective measures of ecosystem condition be based upon a variety of relevant biological attributes but that, to be usable, a comprehensive measure cannot require data for an endless number of system properties. The IBI "represents a synthesis of a dozen distinct hypotheses about the relationship between attributes of biological systems under varying influence from human society" (Karr, 1992, p. 233). The IBI is now widely used in North America and Europe to assess the condition of stream fish communities. Karr's original IBI has been modified to apply the same approach to other organisms, such as stream invertebrates. Carriker (this volume) and Yoder and Rankin apply Karr's approach to assess the conditions of streams and reservoirs managed by the Tennessee Valley Authority and the Ohio Environmental Protection Agency.

CONCLUSIONS

Any efforts to limit human environmental impacts should include two goals. The first should be to minimize the undesirable environmental impact per unit of human activity. The second should be to ensure that the cumulative impact of all human activities is compatible with the persistence of all critical ecosystem conditions and processes. Profound uncertainties will complicate efforts to achieve these goals, but they are appropriate targets.

Environmental performance measures are key tools in identifying opportunities to move toward the first of these goals. Measures of ecosystem condition are vital in charting progress toward the second goal. However, these two sets of tools will be most useful if they can be used to accurately predict the consequences of human activity on affected ecosystems. To achieve this, they must be refined and coordinated such that predictions can be based on the product of the environmental impact per unit of an activity multiplied by the scale of that activity. Such information would be useful for distinguishing acceptable and unacceptable impacts. Users of performance and condition measures should examine the potential for finding or developing relationships between the two types of indicators. Ideally, it will eventually become possible to measure environmental performance in terms that can be related directly to the consequences for affected ecosystems. In essence then, the environmental impact of a given activity could be expressed in units of ecosystem condition. We hope the papers in this volume will foster progress toward that ideal.

REFERENCES

Cairns, J., Jr., P. V. McCormick, and B. R. Niederlehner. 1993. A proposed framework for developing indicators of ecosystem health. Hydrobiologia 263:1– 44.

Daly, H. E., and J. B. Cobb, Jr. 1989. For the Common Good. Boston: Beacon Press.

Hart, D. D. 1994. Building a stronger partnership between ecological research and biological monitoring. Journal of the North American Benthological Society 13:110–116.

Karr, J. R. 1981. Assessment of biotic integrity using fish communities. Fisheries 6:21–27.

Karr, J. R. 1991. Biological integrity: A long-neglected aspect of water resource management. Ecological Applications 1:66–84.

Karr, J. R. 1992. Ecological integrity: Protecting Earth's life support systems. Pp. 223–238 in Ecosystem Health: New Goals for Environmental Management, R. Costanza et al., eds. Washington, D.C.: Island Press.

Lindstedt-Siva, J., L. C. Soileau IV, D. W. Chamberlain, and M. L. Wouch. 1996. Engineering for development in environmentally sensitive areas: Oil operations in a rain forest. Pp. 141–162 in Engineering within Ecological Constraints, P. C. Schulze, ed. Washington, D.C.: National Academy Press.

Schindler, D. W. 1987. Detecting ecosystem responses to anthropogenic stress. Canadian Journal of Fisheries and Aquatic Sciences 44(suppl. 1):6–25.

Schindler, D. W. 1990. Experimental perturbations of whole lakes as tests of hypotheses concerning ecosystem structure and function. Oikos 57:25–41.

Sharpe, R. M. 1995. Another DDT connection. Nature 375:538–539.

Shen, H. W. 1996. Engineering studies based on ecological criteria. Pp. 177–186 in Engineering within Ecological Constraints, P. C. Schulze, ed. Washington, D.C.: National Academy Press.

Solorzano, R., R. deCamino, R. Woodward, J. Tosi, V. Watson, A. Vasquez, C. Villalobos, J. Jimenez, R. Repetto, and W. Cruz. 1991. Accounts Overdue: Natural Resource Depreciation in Costa Rica. Washington, D.C.: World Resources Institute.

Steen, B., and S. O. Ryding. 1992. The EPS Enviro-Accounting Method: An Application of Environmental Accounting Principles for Evaluation and Valuation of Environmental Impact in Production Design. IVL Report B 1080. Göteborg, Sweden: Swedish Environmental Research Institute.

Vitousek, P. M., P. R. Ehrlich, A. H. Ehrlich, and P. A. Matson. 1986. Human appropriation of the products of photosynthesis. BioScience 34:368–373.

NOTE

1. Primary production is the production of plant tissue by photosynthesis. Net primary production is gross primary production (the total amount of carbon fixation by photosynthesis) minus respiration (the consumption of the products of photosynthesis) by the plants.

Life-Cycle Analysis

Measures of Environmental Performance and
Ecosystem Condition. 1999. Pp. 15–28.
Washington, DC: National Academy Press.

Net Energy Expenditure:
A Method for Assessing the Environmental
Impact of Technologies

MARTIN B. HOCKING

Although it is easy to compare the relative environmental merits of bicycling
and driving a car, it is more difficult to use life-cycle inventory methods to compare the merits of paper towels, cloth towels, and a hot-air hand dryer, or of paper
plates and china used in a cafeteria. In other words, it is often difficult to select a
valid approach to compare the environmental advantages of very different technologies, even when they are used to accomplish similar functions.

Various factors such as usefulness, convenience, aesthetics, or the production of waste could be used to weigh the value of parallel technologies; however,
a comparison of the net energy expenditure may be the most useful measure for
judging environmental performance. This is the approach that is explored in this
paper.

Even significant differences in the emission characteristics of competing
technologies can be minimized by the expenditure of some additional energy on
an emission-control function. For example, 2 percent or more of the energy produced by a thermal power station is consumed by its emission-control activities.
Similarly, in sewage treatment plants, the processes used are primarily for emission control and are also, generally, net consumers of energy. Thus, emissions
reduction is one way the environmental merits of competing technologies can be
improved. Energy conservation also can influence the choice of competing
"green" technologies over time.

What is significant about the choice of energy consumption as the primary
factor for comparing technologies? Why not compare the consumption of renewable resources (e.g., biomass) with that of nonrenewable resources (e.g., those of
fossil origin), or the consumption of a large-reserve resource base (e.g., sand,
iron, or salt) with that of a small-reserve resource base (e.g., copper, silver, or

gold), or the use of low-polluting-potential materials (e.g., iron, glass, or cement) with that of high-polluting-potential materials (e.g., lead, mercury, or thallium)? Energy consumption is a valid comparative factor because all of these other distinctions are based ultimately on energy expenditure. Oil, gas, and coal are nonrenewable only in the sense that the energy cost of producing them synthetically from biomass or other carbon sources is 3 to 10 times their current prices. Copper, silver, and gold are more expensive than iron primarily because it takes much more energy to isolate and recover them. Also, these materials are not consumed during use, although some uses may dissipate them. An ounce of gold finely distributed throughout 90 tons of gravel is not worth $400, although at this concentration (about 0.3 mg/kg) it may be worth processing. Until the energy has been expended to recover it, however, it is only worth a very small fraction of $400.

AN OUTLINE OF THE METHOD

Boundaries

To compare technologies using the energy assessment method, appropriate boundaries for the technologies must be defined. During the course of the assessment, additional factors may emerge that require adjusting the boundaries to maintain fairness in the inventory.

Data assembled by Chapman et al. (1974) for automotive transportation and home heating illustrate the importance of appropriate and fair selection of boundaries for a valid comparison of different technologies. For example, the gasoline-powered car has an engine transmission combined efficiency of about 0.2 (Figure 1). The electric car, with a 0.8 battery charge–discharge efficiency and a motor control system–transmission efficiency also of about 0.8 gives a net system efficiency of 0.64, apparently much higher than the gasoline-fueled car.

If the energy-producing facilities for each of these technologies are included in the calculation, however, a very different picture emerges (Figure 1, solid boundaries). Oil production, refining, and delivery systems are estimated to be 0.88 efficient at retaining the energy originally present in the crude oil. Multiplying this by the fractional efficiency of the gasoline-powered car produces an overall efficiency for this system of 0.17. The relatively low efficiency of thermal electricity generation coupled with losses in fuel production and delivery, and electricity lost during transmission, gives an overall efficiency for delivered electricity of about 0.24. Combining this information with the higher efficiency of the electric automobile drive system gives an overall efficiency of 0.15, quite comparable to the overall efficiency of 0.17 calculated for the gasoline-powered system.

This example shows the considerations that enter into establishing boundaries for the conversion of energy into work and demonstrates the kinds of surprises that this exercise can reveal. An examination of appropriate boundaries for

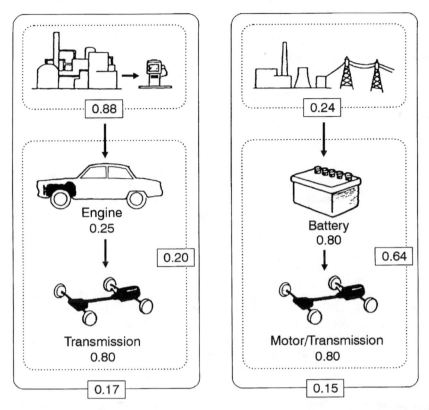

FIGURE 1   Significance of system boundaries on the perception of energy efficiency in automotive transport options (adapted from Chapman et al., 1974). Dotted-line boundaries encompass the fractional efficiencies restricted to automotive options, suggesting that the electric automobile, at 0.64 fractional efficiency, is more efficient than the gasoline-powered automobile. Solid-line boundaries include the efficiencies of the respective energy-delivery subsystems, demonstrating that, overall, the two systems have very similar efficiencies.

the conversion of various sources of energy into heat can also be instructive. Drawing a close boundary for a fossil-fueled boiler for home heating gives a fractional efficiency of about 0.75 (Figure 2). Repeating the process for electric heating gives a near-unity efficiency for the conversion of electrical energy to heat. With close boundaries, it is clear that electric heating is more efficient. If we extend the boundaries to include the respective energy delivery systems, however, the picture changes. The much higher efficiency of the fuel delivery system changes the fractional efficiency of the fueled boiler to 0.66 compared with 0.24 for the electrically heated system. Thus, burning a fuel in the place where the heat is required allows most of the thermal energy to be captured, whereas a large

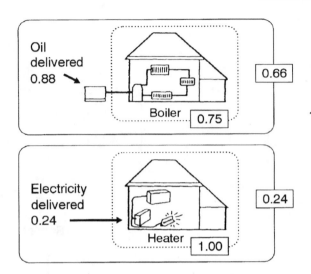

FIGURE 2   Significance of system boundaries on the perception of relative energy efficiencies of home heating options (adapted from Chapman et al., 1974). The relative efficiency within the dotted boundaries suggests that electricity, at 1.00 fractional efficiency,
is more efficient than oil, at 0.75, for heating. However, assuming the larger solid boundaries, inclusive of the respective energy-delivery systems, results in a reversal of the perceived efficiencies: 0.24 for electricity and 0.66 for oil.

fraction of this heat is shed to the environment (i.e., wasted to air or water) during
thermal electricity generation.

Information Sources

Data are available on the energy needed to produce many common materials.
This information can be used as the starting point for an energy analysis. (See,
for example, Berry et al., 1975; Boustead and Hancock, 1979; Gaines, 1981;
Hocking, 1994a,b; Kindler and Nikles, 1979, 1980; Ringwald, 1982.) However,
care must be taken to select the appropriate data. For the production of glassware, for example, published sources give energy requirements ranging from 9.1
to 79.1 J/g of glass, each correct for the particular boundaries and production
conditions considered (Table 1). Researchers may present a range of data rather
than a single number because of variations in the boundaries chosen (which may
or may not be reported), the scale of production, or because more than one energy
source is used. Thus, judicious selection of the most appropriate value is recommended, taking into account as many of the variables that reflect the particular
situation as possible. Similarly, single energy-requirement values should be used

TABLE 1 Range of Energy Costs to Produce Glassware

kJ/g	Remarks
9.1	Glass melting only
9.15	Cornelius furnace,[a] melting only
9.18	Container glass, electricity, 3.06 GJ/mg, × 3[b]
11.8	Container glass and containers
12.6	Container glass, electricity × 3[b]
10.5–17.8	Glass melting
13.3	Container glassmaking, gas fired
14.5	U.K. glass industry, overall, 1982
17.9	1 ton/day electric, from 1,500 kWh/ton,[c] × 3[b]
18.0	Glass containers, 4,534 kWh/ton
19.3	Glassware, melting only, 20 ton/day scale
20.2	U.K. glass industry, overall, 1970
21.0–25.1	Total energy, from raw materials in the ground, generic glass
25.0	Glass bottles, ca. 1980 data
27.7	Glass tableware, includes raw material recovery, transport, plus 40 percent cullet
46.5	Specialty glass, U.K., overall (includes tableware)
79.1	Gas–air firing, melting only

[a]Estimated primary fuel requirement from an electrical requirement of 1 kWh per 1.18 kg of glass.
[b]Electrical energy required is multiplied by 3, assuming a 33 percent efficiency of thermal electricity generation (see text).

SOURCE: Hocking (1994b).

with care. Such values sometimes must be adjusted to better relate the conditions or boundaries of the cited data to the situation under analysis.

When no published data for the product of interest can be located, or when detailed verification is necessary for appropriate selection from a published range, then a detailed inventory of production inputs and outputs is necessary. Doing an inventory is similar to compiling the materials component of a life-cycle inventory (Vigon et al., 1992). The energy components that result from this process can be used to calculate the net energy requirement for that commodity. (See, for example, Hocking, 1991a,b, 1994a,b.) Boundaries must be clearly defined, and any energy conversion factors and equivalencies used in the analysis should be specified so that others may understand the process used and, if necessary, reproduce the results of the analysis.

If a significant proportion of the energy consumed to produce a commodity is from electricity, then the efficiency of conversion of the primary energy source used to generate the electricity has to be considered. Modern fossil-fueled thermal power stations are close to 33 percent efficient; nuclear power generation is slightly more efficient. By comparison, hydroelectric power generation is about 70 percent efficient (Boustead and Hancock, 1979). Thus, with the extensive use of power grids for developed nations, it is possible to estimate a nation's electricity-generating efficiency based on the weighted average of the generating sources used. Grid-generating efficiencies calculated in this way are, for Canada, 57.3 percent; for Norway, 69.9 percent; for the United Kingdom, 35.2 percent; and for the United States, 38.0 percent (International Energy Agency, 1993; United Nations, 1990). Although electricity generation is probably subject to the widest range of production efficiencies, similar considerations apply to the production and delivery of other energy commodities, such as coal, oil, or natural gas (Boustead and Hancock, 1979; Hocking, 1994b).

AN EXAMPLE USING THIS METHOD

A recently completed energy analysis of reusable and disposable cups illustrates the method outlined and the kinds of information that can be obtained. Three reusable cup types—ceramic, glass, and reusable polystyrene—and two disposable cup types—uncoated paper and molded polystyrene foam—are considered.

For the plastic cup types, the boundaries for evaluation included the total energy required—from the extraction of crude oil to production of the final product; for the paper cup, they included the total energy required to produce a finished cup from a standing forest; and for the glass and ceramic cups, they included the energy required to process the raw materials and to produce the finished cups. On the output side, the energy consumed during each cup's life cycle was evaluated up to the point of discard. For the reusable cups, the total operating energy of various commercial dishwashers was calculated. No account was made of the energy or materials required to make a commercial dishwasher, because this energy component per cup-use cycle over the life of the dishwasher will be small relative to the operating energy component.

The energy parameters of interest for each type of cup were compared using equations developed specifically for this purpose (Hocking, 1994b). The published energy requirements for each of the five cup technologies were then collected without regard to their relevance to cups or to whether the reported value was for part of the process or for the whole process (Hocking, 1994b). Detailed examination of these sources permitted selection of appropriate energy data that took into account the specified boundaries (Table 2). Operating details were also obtained for several commercial dishwashers with particular reference to their water, energy, detergent, sanitizer, and rinse agent requirements (Hocking, 1994b). The published values were then converted to common units, kilojoules per gram of material processed, which enabled comparisons within and among technologies.

TABLE 2   Selected Energy Requirements to Make Typical Hot-Drink Cups

Cup Type	Mass Range (g)	Selected Cup (g)	Energy Requirement kJ/g (source)	kJ/cup
Ceramic	227–337	292.3	48.2 (van Eijk et al., 1992)	14,088
Heat-proof glass	166–255	198.6	27.7 (Fenton, 1992)	5,501
Reusable polystyrene	27–109	59.1	106.6 (Fenton, 1992)	6,300
Uncoated paper	6.3–10.2	8.3	66.2 (Hocking, 1991b)	549
Molded polystyrene foam	1.4–2.4	1.9	104.3 (Hocking, 1991b)	198

SOURCE: Hocking (1994b).

Energy of Manufacture

The published energy values for each technology selected for Table 2 represent the best available for the purposes of this study. These values are consistent with the temporal trends in the energy requirements for these technologies established by Fenton (1992), and they are at the upper end of the tabulated range. None are extreme values, however. The highest energy requirement for the fabrication material is for the two types of polystyrene: 104.3 kJ/g for the foamed material and 106.6 kJ/g for the reusable polystyrene varieties. Paper (66.2 kJ/g) and ceramic (48.2 kJ/g) materials required much less energy. Glass required only about one-quarter the energy, or 27.7 kJ/g, needed to make polystyrene.

Arbitrary samples of each cup type (8–9-ounce nominal capacity) manufactured in Canada, China, the United States, and the United Kingdom were weighed, and the median weight of each type was used to calculate the energy of manufacture. On this basis, the very low mass of the molded polystyrene foam cup required the least total energy to produce, 198 kJ/cup, and the ceramic cup the most, at 14,088 kJ/cup.

Energy of Reuse

Most of the electrical energy required for dish washing goes to heat the water, which must be hot for effective cleaning. The electricity required per cycle for the more energy-efficient commercial dishwashers was similar, in the range of 70–83 kJ/cup (Hocking, 1994b). The energy requirements of some models were offset to an extent by reuse of the hot rinse water as the wash water for the next cycle; even so, two of the machines that use this strategy required about 130 kJ/cup per wash. However, electricity is a secondary energy source. To compare dish washing energy with the energy needed to manufacture each cup type, the calculation must include the additional expenditure of primary energy used to generate electricity, as already explained.

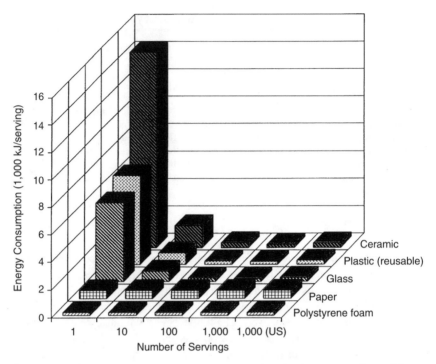

FIGURE 3   Change in energy consumption per serving for three types of reusable cups used only once before washing and two types of disposable cups used only once before discarding. Canadian electricity-generating efficiency was used for the 1-, 10-, 100-, and the first 1,000-serving column. American electricity-generating efficiency was used for the 1,000-serving column labeled US.

Figure 3 illustrates the energy consumption in kilojoules per serving for each of the reusable cup types used only once before washing, calculated using a 184 kJ/cup primary wash energy requirement, an intermediate value representing Canadian generating efficiency. For the disposable cup types used only once before discard, the energy consumption per use is the energy required to manufacture the cup. For a single use, all three types of reusable cups consume more than 10 times the energy per use than do either of the disposable cups. Energy consumption per use of the reusable cups drops to less than that of the paper cup only after about 100 servings. At some point between 100 and 1,000 servings, the energy consumption per use of the reusable cups finally falls to less than that of polystyrene foam disposables.

Another interesting feature emerges from this exercise if one uses the same energy data for an economical commercial dishwasher in combination with the lower average generating efficiency for the United States. In this situation, the median primary energy required to wash a reusable cup is 278 kJ. This is somewhat more than the 198 kJ of primary energy required to make a polystyrene foam cup. In other words, for a single use of both cup types in a country with a low average electrical-generating efficiency, there will be no point at which a reusable cup consumes less energy per use than one use of a polystyrene foam cup (Figure 4). Only if the reusable cup were used twice between washes and the disposable cup used only once before being discarded (not a "level playing field") could any of the reusable cups be a better energy value.

The 278 kJ required to wash a reusable cup with an efficient dishwasher in the United States is less than half of the energy needed to make a paper cup. This means that, compared with a paper cup, a glass cup would use less energy per use after 15 uses, reusable polystyrene after 17 uses, and ceramic after 39 uses. Considering dishwashers of lower energy efficiency, the 340–360 kJ of primary energy per cup used by two of the high-temperature dishwashers in the United States is more than half of the 549 kJ required to make a paper cup. The 549 kJ/cup represents the expenditure of 433 kJ in wood energy, 82 kJ in fossil fuel energy, and 34 kJ in chemicals and processing energy (Hocking, 1994b). Therefore, the

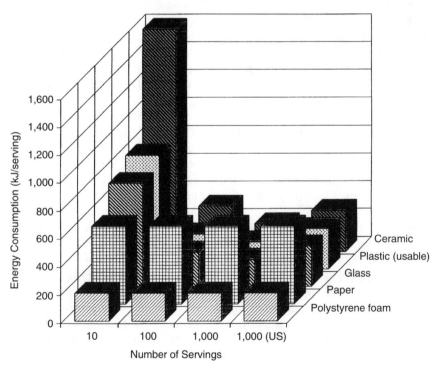

FIGURE 4 As in Figure 3, but with single-serving energies omitted and the energy axis divided by 10. This representation emphasizes the energy-per-serving differences for multiple servings, which appear to be negligible in Figure 3.

per-cup energy use by any dishwasher exceeds the 82 kJ fossil fuel fabrication component (for an 8.3-g cup) of energy required by a paper cup. This means that, using the criterion of fossil fuel consumption, a single use of a paper cup consumes less than a single use of any of the other cup types examined.

If we compare the energy costs per use for the scenario in which the reusable cups are used twice between washes and the disposables are used twice before discard, the picture changes. The reusables must be used between 30 and 2,000 times (per 15–1,000 washes) before they reach break-even per-use energy consumption relative to the disposable cups.

Net energy expenditure of the disposable and reusable polystyrene cup types could be further reduced by recycling or through energy recovery, instead of relying on landfill disposal. The energy reduction achieved would be equivalent to the intrinsic energy content of the material from which they are made, less the energy required to collect these materials for recycling. This would, of course, also shift the energy break-even value. A discarded ceramic cup has no recycle value except as fill, and the energy required to reuse the material in the glass cup as cullet is almost as much as is required in the initial manufacture of glass from raw ingredients (Berry and Makino, 1974; Boustead and Hancock, 1979; Miller, 1983). Thus, no energy is recoverable from ceramic cups, and negligible energy is recoverable from glass cups. But the intrinsic energy recoverable by recycling the reusable polystyrene, the polystyrene foam, and the paper cup types is about 2,364 kJ/cup (59.1 g/cup × 40 kJ/g), 76 kJ/cup (1.9 g/cup × 40 kJ/g), and 166 kJ/cup (8.3 g/cup × 20 kJ/g), respectively (Hocking, 1991b), minus the energy required to collect these materials for recycling.

Trash collection and plastics recycling operations have been estimated to have energy costs of 0.28–0.40 kJ/g and 26.7 kJ/g, respectively (Berry and Makino, 1974). If there is a similar energy cost to recycle paper, recycling the material of the three cup types back into cups results in net energy costs of 1,599 kJ/cup [(59.1 g/cup × 0.36 kJ/g collection) + (59.1 g/cup × 26.7 kJ/g reprocessing)] for the reusable polystyrene cup, 51 kJ/cup [(1.9 g/cup × 0.36 kJ/g) + (1.9 g/cup × 26.7 kJ/g)] for the polystyrene foam cup, and 222 kJ/cup [(8.3 g/cup x 0.36 kJ/g) + (8.3 g/cup × 26.7 kJ/g)] for the paper cup. For this hypothetical scenario, then, recycling these materials entails energy costs equal to about one-fourth those involved in using virgin material for both plastic cup types and two-fifths of that required if virgin material is used for the paper cup. The larger fractional benefit of recycling plastic cups is due solely to the larger intrinsic energy content of polystyrene. This illustrates an important general relationship between materials and recycling: The higher the intrinsic energy content of the material, the higher the potential energy savings that is obtained by recycling rather than using virgin material. It should be pointed out, however, that direct recycling of materials such as these back into food service is not permitted by law, for public safety reasons, without at least a layer of virgin material placed on the food-contact side of the container. In this example, if energy costs of recycling (collection/sorting, transporting, reprocessing) can be kept low and the grade of end use of the recycled material kept high to maintain the high intrinsic energy content of these materials, then in terms of energy consumption, recycling is an attractive option. Otherwise, recycling is a less attractive option, and an energy recovery strategy is preferable.

COMMENTARY ON THE EXAMPLE

The range of correct values for some of the energy inputs considered in this analysis was particularly broad for the manufacture of ceramic and glass cups. But the analysis revealed that the break-even points are not as sensitive to changes in this parameter (the fabrication energy of the reusable cup types) as they are to the energy required for washing and sanitizing the reusable cups or fabricating the disposable cups. This analysis also demonstrated that the high fabrication energy required for the reusable cups became unimportant over enough uses, say 500 or more, compared with the energy required to wash and sanitize them for reuse. It also revealed that the wash energy alone is as much or more than that required to make a polystyrene foam cup in the United States and more than half that required to make a paper cup. Therefore, from an energy standpoint, use of disposable cups is appropriate, especially in situations where the return and reuse rate of the reusable cups is likely to be low.

Many people may prefer to use a ceramic, glass, or reusable polystyrene cup rather than a disposable cup. It is difficult to determine how much weight should be given to this "personal-preference" factor. However, this comparison of energy consumption demonstrates that it is environmentally reasonable to use the disposable cup types when the return rate of reusable cups is likely to be low, or for situations of one-time use such as for large parties, because the energy required for manufacture of the disposable cup types is less than, or very close to, the energy needed to clean a reusable cup.

Finally, this analysis confirms that there is environmental merit in having a diversity of cup types available depending on type of use. In certain circumstances, disposable cups may actually use fewer resources and cause less environmental impact than ceramic, china, or glass cups.

CONCLUDING NOTES

In an energy-based analysis of relative ecological merit, it is crucial to define clearly the boundaries to be used. The analyst must also be prepared to adjust these boundaries as the data are gathered if the initial limits to the system appear inappropriate.

Energy costs of raw materials will vary with location, so the area where the technology is being studied has to be specified. Correction factors may have to be used to adjust the energy components used in the analysis, but they should be

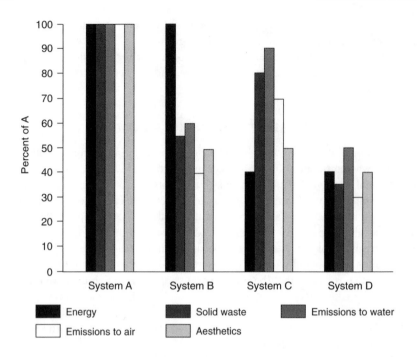

FIGURE 5   Representations of energy consumption, waste-stream discharges, and aesthetic preferences for four hypothetical technology systems.

specified and the rationale for their selection explained. Spelling out the specifications of an analysis makes it possible for others to repeat it, which lends validity and credibility to the results.

It is possible to apply sensitivity tests to an energy analysis by decreasing or increasing key energy terms and observing the effect on the results. (See, for example, Hocking, 1994b.) In this way, it is possible to identify the factors most important for decreasing the energy consumption of a technology.   In the cup example, sensitivity tests demonstrated that the significant factors were the energy requirement for washing reusable cups and the fabrication energy for the disposable cups.

Although energy analysis can be used to compare widely differing technologies and is a useful tool for the quantitative determination of resource expenditure, it contributes little information on the relative emission characteristics of alternate technologies.   The primary energy sources, whether coal-fired or gas-fired boilers, or hydroelectric, will have different emission characteristics, but the energy assessment process itself does not give this information.

Inventory versus Analysis

A comprehensive picture of a technology's environmental impact requires both an energy assessment and a life-cycle inventory. The analyst must catalog the available energy consumption data and assemble the data relating to solid-waste and water and air emissions. Even with this information, it may be difficult to assess the relative merits of two or more sets of inventory data. Four hypothetical scenarios demonstrate this difficulty (Figure 5). System B has the same energy consumption as System A but produces significantly lower quantities of all three wastes. To compare System B with System C, which has higher volumes of all three waste streams but much lower energy consumption, requires a judgment call or an "eco-points" rating system of some kind. Does the fact that System C has a lower energy consumption offset the higher emission rates? System D, which has both the lowest energy consumption and the lowest emission rates, is most readily rated the best of the four. System B or C could be considered better than A and worse than D, but some kind of aesthetic, ecological, or resource rating system is needed in order to decide.

The scenarios presented in Figure 5 are relatively simple. Even so, the products are difficult to rank. If we increase the level of detail in each of the waste streams to the point of quantifying the particular components of the solid-waste streams (e.g., suspended solids, dissolved solids, biochemical oxygen demand [BOD], for emissions to water; particulate matter, sulfur dioxide, carbon monoxide for emissions to air) the difficulty of ranking technologies increases enormously. How does a technology with high sulfur dioxide emissions and low suspended solids and BOD rate relative to one in which negligible sulfur dioxide is discharged and the aqueous waste stream is high in suspended solids and BOD? Is there a way to assess or weigh equivalent impacts of an air emission and a water emission? For air emissions, such assessments can be based on human toxicity or photochemical-smog-forming potential (Hocking, 1985), but ratings of pollutant categories again require judgment calls (i.e., become subjective).

Finally, there is the difficulty of weighing the significance of the aesthetic, or the personal-preference, factor in technology choices. In Figure 5, System A, which could represent one of the reusable cups, has a high aesthetic rating but also high energy costs and emission loadings. System D, which could represent a disposable cup, has a low aesthetic rating but ecologically favorable low energy costs and low emission loadings. Again, it is a difficult choice to make. Perhaps the best solution is to continue to provide a diversity of cup technologies for the diverse applications to be met. In this way, a reusable cup that begins to become energy competitive after enough uses can fulfill both a reasonable resource expenditure and aesthetic needs.

Alternatively, in situations where little or no reuse is likely, selecting the appropriate disposable cup and accommodating consumer preferences (i.e., insulated for hot drinks or noninsulated/clear for cold drinks) would be the best choice.

ACKNOWLEDGMENTS

The author is grateful to the University of Victoria for support of this work and to the many individuals in companies whose readiness to provide information made the reported analyses possible. They are acknowledged wherever feasible in the published full accounts of this work.

REFERENCES

Berry, R. S., and H. Makino. 1974. Energy thrift in packaging and marketing. Technology Review 76(4):32–43.

Berry, R. S., T. V. Long II, and H. Makino. 1975. Energy budgets: 5. An international comparison of polymers and their alternatives. Energy Policy 3(2):144–155.

Boustead, I., and G. F. Hancock. 1979. Handbook of Industrial Energy Analysis. New York: John Wiley & Sons.

Chapman, P. F., G. Leach, and M. Slesser. 1974. The energy cost of fuels. Energy Policy 2(3):231–243.

Fenton, R. 1992. The Winnipeg Packaging Project: Report No. 2. Comparison of Coffee Cups. Winnipeg, Manitoba: The University of Winnipeg.

Gaines, L. L. 1981. Energy and Materials Use in the Production and Recycling of Consumer Goods Packaging. Report ANL/CNSV-TM-58. Argonne, Ill.: Argonne National Laboratories.

Hocking, M. B. 1985. Modern Chemical Technology and Emission Control. New York: Springer-Verlag.

Hocking, M. B. 1991a. Paper versus polystyrene: A complex choice. Science 251:504–505.

Hocking, M. B. 1991b. Relative merits of polystyrene foam and paper in hot drink cups: Implications for packaging. Environmental Management 15:731–747.

Hocking, M. B. 1994a. Disposable cups have eco merit. Nature 369:107.

Hocking, M. B. 1994b. Reusable and disposable cups: An energy-based evaluation. Environmental Management 18:889–899.

International Energy Agency, Organization for Economic Co-operation and Development (OECD). 1993. Energy Statistics of OECD Countries, 1990–1991. Paris: OECD.

Kindler, H., and A. Nikles. 1979. Energiebedarf bei der Herstellung und Verarbeitung von Kunststoffen. Chemie-Ingenieur-Technik 51(11):1125–1127.

Kindler, H., and A. Nikles. 1980. Energieufwand zur Herstellung von Werkstoffen: Berechnungsgrundsätze und Energieäquivalenzwerte von Kunststoffen. Kunstoffe 70 (12):802–807.

Miller, R. K., ed. 1983. Energy Conservation and Utilization in the Glass Industry. Atlanta, Ga.: Fairmont Press Inc.

Ringwald, R. M. 1982. Energy and the chemical industry. Chemical & Industry 9:281–286.

United Nations. 1990. World Resources, 1990–1991, A Report by the World Resources Institute and the United Nations Environment Programme. Oxford, England: Oxford University Press.

van Eijk, J., J. W. Nieuwenhuis, C. W. Post, and J. H. de Zeeuw. 1992. Reusable versus Disposable. A Comparison of the Environmental Impact of Polystyrene, Paper/Cardboard, and Porcelain Crockery. Zoetermeer, Netherlands: Ministry of Housing, Physical Planning and Environment.

Vigon, B. W., D. A. Tolle, W. Cornaby, H. C. Latham, C. L. Harrison, T. L. Boguski, R. G. Hunt, and J. D. Sellers. 1992. Product Life Cycle Assessment: Inventory Guidelines and Principles. Cincinnati, Ohio: U.S. Environmental Protection Agency.

Measures of Environmental Performance and
Ecosystem Condition. 1999. Pp. 29–41.
Washington, DC: National Academy Press.

Life-Cycle Analysis:
The Role of Evaluation and Strategy

FRANK R. FIELD III AND JOHN R. EHRENFELD

Life-cycle analysis (LCA) has become one of the most actively considered
techniques for the study and analysis of strategies to meet environmental challenges. The strengths of LCAs derive from their roots in traditional engineering
and process analysis. Also vital is the technique's recognition that the consequences of changes in technological undertakings may extend far beyond the
immediate, or local, environment. A technological process or a change in process
can produce a range of consequences whose impacts can only be perceived when
this entire range is taken into consideration. The application of LCA promises to
change the treatment of environmental considerations within the larger concerns
of modern technological society. However, as the technique becomes more popu-
lar, it is becoming clear that some of the problems LCA is expected to solve lie
outside its practical and conceptual boundaries.

Potential users of this technique span a wide spectrum of interests. Process
and product developers view LCA as a way to incorporate environmental considerations into their design process, making it possible to anticipate and avoid potential pitfalls. Consumers and consumer interest groups see LCA as a way to
better inform the customer of the relative environmental impact of alternative
products, hoping to bring market pressures to bear on producers. Finally, regulators and policy makers see LCA as a tool to guide the development of environmental policy and also provide a mechanism to enforce legislative objectives.

The development of LCA arose largely from the need for tools that take
account of the growing social importance of environmental objectives. The market, the principal way in which consumer interests are translated into technological action, currently does not supply consumers with environmentally relevant
information. The complexity of the modern industrial economy makes it difficult

to know how any individual action affects the environment. LCA is being developed to produce a framework within which this information can be collected, refined, and acted on.

However, analysis of any kind is limited in its ability to resolve complex problems, particularly when an action has consequences that advance some objectives while hindering others. Under these conditions, the choice among alternatives must incorporate not only analytical elements, but strategic ones as well. LCA is well suited to supplying the former but not the latter.

LIFE-CYCLE ANALYSIS

The basic objective of LCA is to guide decision makers, whether consumers, industrialists, or government policy makers, in devising or selecting actions that will minimize environmental impacts while furthering other objectives. Decision makers must use this tool in concert with traditional criteria for selecting one action over another, including economic, engineering, and social goals.

The life-cycle paradigm requires the consideration not only of the immediate impacts of a product or process choice, but also of the products and processes that gave rise to that choice and of those that occur in response to it. This view reflects the notion that "industrial ecosystems," like natural ecosystems, are vast networks of interconnected activities. In such networks, the size of a particular change does not necessarily indicate the scope of its effect, and care must be taken to avoid changes that maximize local benefits at the expense of global effects.

LCA is a three-step process:

• inventory analysis, or the identification and quantification of energy and resource use and environmental releases to air, water, and land;
• impact analysis, or the technical qualitative and quantitative characterization and assessment of the consequences of resource use and environmental releases for the environment; and
• improvement analysis, or the evaluation and implementation of opportunities to reduce environmental burdens (Vigon et al., 1993).

The three stages of LCA reflect classical technical decision-making procedures. In each case, a control volume is identified. Resource flows into the control volume and waste emissions from the control volume are then measured. The next step is to determine the relationship between these resource and waste fluxes and the underlying scientific and technological principles. Finally, the problem is resolved based on the insight gained from these principles and the objectives of the analyst.

Much of the focus on LCA has been on how and why it is used. Organizations such as the Society of Environmental Toxicology and Chemistry and the

U.S. Environmental Protection Agency (EPA) have worked to develop a complete set of procedures to use in collecting and organizing the information that must be developed in the course of an LCA (Fava et al., 1990; Vigon et al., 1993). However, many observers remain uncertain about what to do with this information once it is collected. Expressed simply, LCA is a tool for enhancing positive environmental impacts.

Unfortunately, except in the simplest of situations, it is extremely difficult to determine how this general objective informs specific problems—a fact that increasing numbers of LCA practitioners recognize. This difficulty arises from several sources. The most apparent of these is the imperfect understanding of the relationship between releases to the environment and environmental damage, particularly when many such releases must be considered together. However, this limitation apparently has not inhibited the development and application of LCA methodology.

Improvement analysis has proved to be the most complicated aspect of LCA. Improvement analysis assumes that it is possible to discern the best action from a set of possible options. In simple cases, it may be possible to find an action that reduces all impacts on the environment. More often, the best course of action requires an assessment of the relative importance of each of a number of possible consequences. These relative importances reflect the strategic objectives that underlie the problem being considered rather than the results of any purely analytical evaluation. Because of this distinction, substantial hurdles must be overcome before LCA can be applied to broad questions of industrial and social policy.

REVIEW OF VALUATION CONCEPTS

This difficulty can be best understood by considering the general problem of valuation (see Goicoechea et al., 1982). Figure 1 depicts a hypothetical set of potential alternatives, each of which has (for the sake of illustration) only two characteristic environmental impacts, A and B. Assuming that only one alternative can be chosen, and that the objective is to reduce environmental impact, which alternative should be implemented?

It is easy to reject alternatives 2 and 3, because other choices exist (alternative 7, for example) that reduce impact A and impact B. Alternatives 2 and 3 are members of what is known as the *dominated set* of alternatives; they are clearly inferior to others. In environmental terms, rejection of the dominated alternatives is an expression of the so-called precautionary principle, which favors taking any action that unequivocally reduces all environmental impact. Similarly, an LCA that showed a facility operating at point 3 would lead to the implementation of an alternative (5 or 6) that reduces all impacts on the environment.

The difficulty arises when a choice must be made among alternatives on the lower edge of the frontier. Which one of these is the best way to operate? Decision analysis refers to these remaining points as the set of *nondominated alterna-*

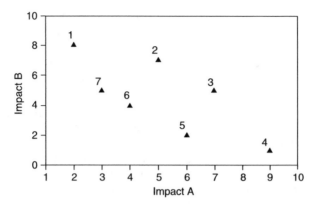

FIGURE 1   Hypothetical set of potential alternative technologies or procedures, each of which has only two characteristic environmental impacts, A and B. Decision analysis refers to the points 1, 4, 5, 6, and 7 as the set of nondominated alternatives. No member of the set is better than the others in all respects.

tives, meaning that no member of the set is better than the others in all respects. Rather, some are better in one or more aspect but worse in at least one other.

How best to select among members of the nondominated set of alternatives is one of the central questions of decision analysis and is frequently referred to as multiple-objective decision making. In multiple-objective decision making, there is no generally applicable rationale for selecting one alternative over the other; rather, the choice requires taking into account strategies and priorities. As Figure 1 shows, the only supportable reason for selecting alternative 6 over 5 is that reducing impact A is more important than reducing impact B.

In decision analysis, the simplest method for selecting from the non-dominated set is to identify specific limits that either must be met or cannot be exceeded. When such constraints are imposed, the set of alternatives can be reduced, as shown in Figure 2. This approach mirrors the traditional command-and-control environmental regulatory model. However, it has important limitations when applied to environmental impact and LCA. The most obvious is that it is almost impossible to establish these limits for every potential impact. In addition, the figure illustrates a more subtle, and potentially more troubling, limitation. Note that alternative 7 is rejected in favor of alternative 6, even though the differences in impact B between the two are relatively small, whereas the differences in terms of A are relatively large. Is it really worthwhile to sacrifice the potential gains in terms of A that alternative 7 represents merely because it barely fails to meet the fixed limit on impact B?

The use of *value functions* overcomes this limitation of simple constraint-setting (or screening) methods of decision making. These functions represent

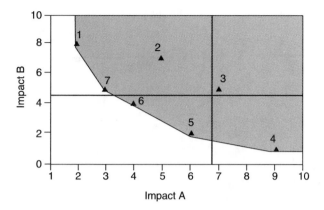

FIGURE 2 Hypothetical set as in Figure 1 with the addition of a line connecting the set of nondominated alternatives and vertical and horizontal lines denoting hypothetical maximum limits (e.g., regulatory limits) for impacts A and B.

preferences among the several attributes that form the basis for the decision (in this case, impacts A and B). The simplest form of a value function is represented in the top panel of Figure 3, the linear index. Essentially, this index estimates a measure of value by constructing a weighted average of the (two) criteria. The option yielding the best average value is selected. Alternatively, a nonlinear value function can also be constructed, as shown in the bottom panel. This value function can represent such observed preference behavior as saturation (i.e., attaining better levels of one attribute reduces the incremental value of further improvement) and variable rates of transformation among attributes.

The linear index method is directly analogous to the concept of *monetization*, the transformation of attributes into their dollar equivalents. (See, for example, the Swedish Environmental Priority Strategies (EPS) system [Steen and Ryding, 1992].) The straight line depicted in Figure 3 can then be thought of as a "budget" for environmental damage. Alternatively, the nonlinear preference function methods directly represent the consumer economist's classical notion of cardinal utility, where the curved line represents a line of constant utility. The curved line in Figure 3 then represents all combinations of environmental damage from A and B that leave the observer equally well (or poorly) off.

As the figures demonstrate, both of these value functions establish that an alternative exists that is demonstrably the "best"; the point of tangency between the line or curve of constant value and the gray area is the alternative that yields the best combination of characteristics. Although establishing a best alternative in the real world requires considering a much larger set of attributes, the conceptual basis remains the same.

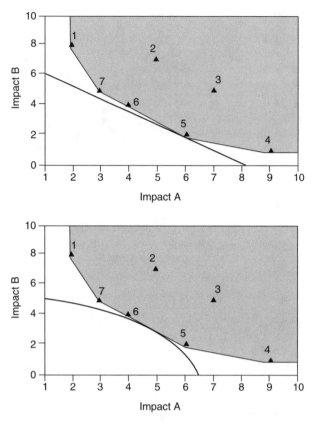

FIGURE 3 Hypothetical set as in Figure 2 with maximum-limit lines removed and lines added that represent alternative linear and nonlinear value functions. Each line connects points of equal value based on combinations of the two impacts.

ENVIRONMENTAL IMPACT ASSESSMENT IN
LIFE-CYCLE ANALYSIS

This review of decision analysis suggests that users of LCA will face two clear-cut classes of problems in the final improvement-analysis stage of LCA where environmental impacts are assessed. In one class, the analyst will be confronted with the choice among several alternatives, one of which clearly dominates. This situation is analogous to choosing between an alternative that lies within the gray area in Figure 3 and another that lies on the lower edge of that area; that is, it is a choice between a nondominated and a dominated alternative. In this case, assuming the LCA treats the complete scope of environmental consequences, it will have revealed that one alternative has better environmental performance in all aspects. For the rational decision maker facing this class of problem, LCA will have unquestionably made the choice easier. The second class of problems, however, will be much more difficult to resolve. In this situation, the two possible alternatives both lie on the lower edge of the gray area; that is, the choice will be between two nondominated alternatives. The analyst therefore cannot resolve the problem without the application of some value function, which itself must represent the strategic interests of the community that the analyst is attempting to serve.

In these cases, establishing the relevant value functions will be a crucial element of the improvement analysis. Individuals and probably many firms can develop these functions using a variety of techniques and appropriately structuring the decision problem (Dyer and Forman, 1992; Keeney and Raiffa, 1976). However, substantial complexities are associated with a wider application.

The Swedish EPS system illustrates both the potential and the limitations of valuation methods when applied in such complex situations. This system, developed specifically with LCA in mind, employs monetization to establish the value of alternatives. Its application is currently being evaluated and endorsed by the Volvo Car Corp., among other companies. (For a complete treatment of this method, consult the references at the end of this paper.)

LINEAR VALUATION: THE ENVIRONMENTAL PRIORITY STRATEGIES SYSTEM

The EPS system is under development by the Swedish Environmental Research Institute, Chalmers Institute of Technology, and the Federation of Swedish Industries (Steen and Ryding, 1992). The system is designed as a tool for evaluating the ecological consequences of alternative activities or processes and ultimately for generating a value for the various changes to the environment induced by these activities.

The EPS system is specifically constructed to associate an environmental load with individual activities or processes, based on materials consumed or processed per unit. For example, EPS might associate X number of environmental load units (ELUs) per kilogram of steel produced and Y ELU per kilogram of steel components stamped. Thus, the environmental load of stamping a 5-kg automobile component, requiring 5.3 kg of steel, would be $5.3 X + 5 Y$. This result could then be compared with the load associated with a different process or the use of a different material. The interesting questions are: How are these environmental loads established? and What do they mean?

Based on the environmental objectives of the Swedish Parliament, the EPS system relates all of the physical consequences of the processes under consideration to their impact on five environmental safeguard subjects: biodiversity, production (growth and reproduction of nonhuman organisms), human health, resources, and aesthetic values. Because a process may affect any one safeguard subject in several forms, EPS allows for the individual consideration of each of these consequences, called unit effects. Two criteria are applied when establishing which impacts will become unit effects: How important the impact is on the sustainability of the environment; and is it possible to establish a quantitative value for that impact within traditional economic grounds. Examples of unit effects for human health include mortality due to increased frequency of cancer, mortality due to increased maximum temperatures, and decreases in food production (and hence increased incidence of starvation) due to global warming.

Once the individual unit effects are established, their values must be determined. This is accomplished by expressing each unit effect in terms of its economic worth and associated risk factors. Formally, the value of each unit effect is equal to the product of five factors, F1 through F5. F1 is a monetary measure of the total cost of avoiding the unit effect. The extent of the affected area (F2), how frequently the unit effect occurs in the affected area (F3), and the duration of the unit effect (F4) represent risk factors similar to those used in toxicological risk evaluations. F5 is a normalizing factor, constructed so that the product F1 × F5 equals the cost of avoiding the unit effect that would arise through the use or production of one kilogram of material. The product of all five factors yields the contribution of a particular unit effect to environmental load. Summing the values of each unit effect yields the environmental load index (ELI) in units of environmental load per unit of material consumed or processed (ELU per kilogram), as summarized in Figure 4. Because these unit effects were specified according to their relevance to the five safeguard subjects, the ELI represents the total environmental load (or impact) for all five safeguard subjects.

For example, consider Table 1, which illustrates how to estimate the ELI for the release of carbon monoxide (CO) to the air. The second and third columns of data demonstrate how the impact of two specific unit effects, nuisance and morbidity, are incorporated into the overall ELI for a CO release to air. Based on a variety of studies, the value of excess nuisance and morbidity are estimated at 10^2 and 10^5 ELU/person-year, respectively. (Note that according to the definition of F1, these values are the estimated costs, in ELUs, of avoiding these unit effects.) Furthermore, the incidence of these impacts is estimated for the world urban population, assuming that hazardous levels of CO occur only 10 percent of the time, and that 10 and 0.1 percent, respectively, of the exposed population is affected at the nuisance or morbidity level. Finally, given that 1,600 million metric tons of CO are already being released, the incremental effect of one additional kilogram released is 1/1,600,000,000,000, the F5 term. These terms, F1–F5, are multiplied together and then summed over all unit effects to develop the ELI for CO release to the atmosphere, in ELUs per kilogram released. With this number, any life-cycle data that reveal the release of some amount of CO can be valued by multiplying that release by the ELI. The EPS system is designed to develop ELIs for all releases, as well as for all human activities that consume resources, so that the relative ELUs for any two life-cycle inventories can be computed and compared.

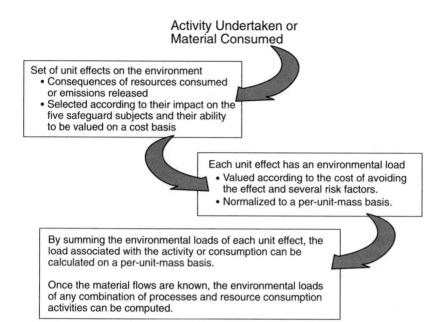

Activity Undertaken or Material Consumed

Set of unit effects on the environment
• Consequences of resources consumed or emissions released
• Selected according to their impact on the five safeguard subjects and their ability to be valued on a cost basis

Each unit effect has an environmental load
• Valued according to the cost of avoiding the effect and several risk factors.
• Normalized to a per-unit-mass basis.

By summing the environmental loads of each unit effect, the load associated with the activity or consumption can be calculated on a per-unit-mass basis.

Once the material flows are known, the environmental loads of any combination of processes and resource consumption activities can be computed.

FIGURE 4   Flow chart summarizing the procedures used in the Environmental Priority Strategies (EPS) system. SOURCE: Steen and Ryding (1992).

VALUATION ISSUES RAISED BY THE ENVIRONMENTAL PRIORITY STRATEGIES SYSTEM

This formulation of valuation raises important questions of scientific feasibility. Indeed, it is debatable whether it is possible to characterize fully the unit effects of every process or activity that might be developed. However, the crucial valuation questions arise from two other aspects of this scheme: the nature of the economic measures used in calculating the cost of avoiding a unit effect and the assumption that the value of the total environmental impact of an action (the environmental load) is equal to the sum of each individual environmental load weighted by the size of each unit effect.

The first of these questions relates to the distinction between cost and worth. Although the theory of competitive markets argues that prices *are* an object's worth, the theory rests on assumptions that are difficult to support in the case of the environment. The first problem is that perfect markets assume that perfect information is available to all participants, which clearly is not the case. Furthermore, the theory of markets routinely discusses "consumer surplus," which can roughly be determined as the difference in the prevailing market price and the

TABLE 1 Calculation of Environmental Load Index (ELI) for 1 kg of CO Released to the Air

Unit Effect	Nuisance	Morbidity	CO_2 Effect	Oxidant Effects
Safeguard subject	Human health	Human health	All 5	All 5
Impact measure	CO concentration	CO concentration	CO_2 equivalents	Ethylene equivalents
F1, value	100^a	$100,000^a$	0.08887^b	0.0005^b
F2, persons affected	$750,000,000^c$	$750,000,000^c$	1	1
F3, frequency or intensity	0.1^d	0.001^d	3^e	3^e
F4, duration	0.01^f	0.01^f	1	1
F5, contribution to total effect	6×10^{-13g}	6×10^{-13g}	1	1
ELI contribution	0.000045	0.00045	0.266202	0.0015
ELI for 1 kg of CO released to the air		0.27		

[a]F1 values for the first two unit effects reflect the assessment that the value of "moderate nuisance" is 10^2 ELU/person-year and that of "painful morbidity and/or severe suffering" is 10^5 ELU/person-year.

[b]Because CO has impacts similar to those of CO_2 and ethylene, the ELIs for these two species are given in the last two columns of the F1 row. By definition, the ELI already aggregates the impact over all pertinent safeguard subjects.

[c]The F2 value is the scope of the effect; in this case, an estimated one-third of the global urban population (~2,280 million) is exposed to excessive CO concentrations 1 percent of the time.

[d]The F3 values represent World Health Organization estimates that 10 and 0.1 percent of those exposed to concentrations of CO above recommended levels are affected at the nuisance or morbidity level, respectively.

[e]The magnitude of the impact of CO is estimated at three times the impact of CO_2 and ethylene, so the F3 term is 3. Other terms are set to 1, because these other effects have already been captured in the ELIs of the equivalent chemicals

[f]The F4 values represent estimates that critical levels of exposure to CO are experienced 1 percent of the time.

[g]The F5 values represent the incremental effect of 1 kg of CO released. Because global human-caused releases of CO are estimated at 1,600 million metric tons/yr, the incremental impact of 1 kg released is 6.25×10^{-13}.

SOURCE: Steen and Ryding (1992).

higher price that some consumers would have been willing to pay (recall that demand curves slope downward). Finally, there is the critical question of how to establish these costs and prices when markets do not exist. Although litigators are prepared to place a value on wrongful death or pain and suffering during a civil suit, no markets exist for pain, clean air, or future well-being. Generally, most environmental attributes are external to markets; many of the classical examples of market externalities are based on environmental issues.

Where markets exist, EPS uses market prices to establish the costs of avoidance. Where market prices do not exist, EPS relies on two alternatives. If government funds are allocated to resolve specific problems (for example, to protect a particular species), these funds are normalized and extrapolated to obtain a cost figure; in this case, the value of maintaining biodiversity is established by normalizing the government's annual budget for species protection. If no funds have been allocated, then the method of contingent valuation is employed. This method (or set of methods) is based on direct inquiries of representative populations to determine their willingness to pay to avoid specific effects. As might be expected, this last approach to establishing the appropriate costs of avoidance is controversial because it is hard, both conceptually and practically, to design questions that demonstrably extract the correct measure of value.

The second of these valuation questions reflects the fact that the mathematical structure of the value function is a consequence of critical assumptions about the nature of the subject's preferences. The valuation used in the EPS system is an example of a linear, additive preference structure. Each unit effect is reduced to a monetary value, normalized for risk and exposure and for material quantity. Thereafter, the net impact of each increment in unit effect is the same, regardless of both how large the effect is and the size of any other unit effect. Although such value functions are simple to represent and employ (i.e., as linear combinations of linear functions), they are not the most accurate, general-purpose formulation of value functions for environmental impact. Although the appropriate form of the value function may be linear, EPS does not explicitly make this assumption. Rather, the linearity of the EPS valuation is based on the assumption that, because monetization reduces all effects to a common metric, the resulting metrics should be additive. In fact, most individuals do not even exhibit linear preferences for money, much less for more subjective attributes. (For example, most individuals would consider paying $0.50 to play a game offering a 50:50 chance of winning $1.00, while rejecting out of hand paying $5,000 for a 50:50 chance of winning $10,000.) In practice, preferences usually reflect nonlinearities both in individual effects and in substitution between effects.

A THIRD LIMITATION: INDIVIDUALS VERSUS GROUPS

Viewing money as a measure of value and calculating linear additive preferences are not necessarily unworkable approaches when considering the development of value functions for the environment. Although difficult, it may be possible for someone to establish the dollar value that exactly offsets a particular unit effect. Similarly, linear additive preferences may be able to model the behavior of an individual over a restricted range. However, it is impossible to state that every individual in the affected population will agree to the same dollar value or the same summing of preferences for environmental considerations. If individuals cannot agree on the value or the structure of their preferences, then no single value function can be constructed to represent their wants.

Conceptually, value functions are based on the notion of individual preference, reflecting strategic objectives. Value functions assume that, given two alternatives, the individual decision maker can say one of two things about them: one alternative is better than the other, or both alternatives are equally good.

The assumptions underlying the concept of value functions are particularly weak when considering the problem of establishing group preferences for environmental attributes. There are two reasons for this. First, to choose between two or more alternatives, the implications of the choice must be fully understood. Otherwise, the choice is meaningless and essentially random. When experts cannot establish the incremental effects of the potential changes in environmental release and resource consumption of two or more alternatives, it is virtually impossible to expect these experts, much less the public at large, to say that one is preferable to the other. Second, even if all the implications of each choice were characterized to the complete satisfaction of all members of the group, individuals still do not have a consistent set of objectives when confronted with environmental choices. For example, some might believe that preventing global warming is more important than reducing urban air pollution, whereas others might think that neither of these objectives is as important as maintaining and improving human health. This lack of a consistent set of priorities in the environmental area essentially eliminates the possibility of constructing a useful value function.

Although the EPS system is a commendable attempt at simplifying the enormous detail of inventory data, the system's developers have pointed out that it is based on their subjective value judgments, which are not necessarily supportable in all situations worldwide. The goals set out by the Society of Environmental Toxicology and Chemistry and the EPA for improvement analysis based on life-cycle inventories are laudable, but they can only be realized by some type of consensus on the value of avoiding environmental degradation.

This suggests that achieving the final stage of LCA will require the development of a basis for devising (and revising) this consensus. In the absence of a common strategic objective, it will be impossible to use LCA to designate ways to achieve environmental improvement beyond straightforward strategies for pollution prevention or the use of precautionary principles. A strategic consensus is required to trade off competing environmental, economic, and engineering goals.

SUMMARY

LCA is a technique that has already shown great promise for improving our understanding of the wider implications and relationships that must be taken into consideration when incorporating environmental concerns into technical decision making. As these concepts diffuse into industrial and technical decision making, LCA will enable industry and government to find ways to both increase efficiency and reduce harm to the environment.

However, practitioners and proponents must guard against using LCA to determine "best" modes of action when the consequences of the alternatives expose conflicting objectives and values within the group of decision makers. In these cases, no amount of analysis will directly resolve the conflict. Rather, the role of LCA should be to clearly articulate the consequences of each alternative and to provide a framework for the necessary negotiations.

REFERENCES

Dyer, R. F., and E. H. Forman. 1992. Group decision support with the analytic hierarchy process. Decision Support Systems 8:99–124.

Fava, J. A., R. Denison, B. Jones, M. Curran, B. Vigon, S. Selke, and J. Barnum, eds. 1990. A Technical Framework for Life-Cycle Assessments: Workshop Report. Society for Environmental Toxicology and Chemistry, Foundation for Environmental Education. Smuggler's Notch, Vt., August 18–23, 1990.

Goicoechea, A., D. R. Hansen, and L. Duckstein. 1982. Multiobjective Decision Analysis with Engineering and Business Applications. New York: John Wiley & Sons.

Keeney, R. L., and H. Raiffa. 1976. Decisions with Multiple Objectives: Preferences and Value Tradeoffs. New York: John Wiley & Sons.

Steen, B., and S. O. Ryding. 1992. The EPS Enviro-Accounting Method: An Application of Environmental Accounting Principles for Evaluation and Valuation of Environmental Impact in Production Design. IVL Report B 1080. Göteborg, Sweden: Swedish Environmental Research Institute.

Vigon, B. W., D. A. Tolle, B. W. Cornaby, H. C. Lathan, C. L. Harrison, T. L. Boguski, R. G. Hunt, and J. D. Sellers. 1993. Life-Cycle Impact Assessment: Inventory Guidelines and Principles. EPA/600/R-95/245. Washington, D.C.: U.S. Environmental Protection Agency.

Measures of Environmental Performance and
Ecosystem Condition. 1999. Pp. 42–62.
Washington, DC: National Academy Press.

Defining the Environmentally
Responsible Facility[*]

BRADEN R. ALLENBY AND THOMAS E. GRAEDEL

Traditionally, environmental concerns and subsequent regulation have focused on perturbations that were local in both time and space (e.g., individual waste disposal sites, specific airsheds or watersheds). The desire was to clean up the air over Los Angeles or to make the Hudson River clean enough to support fisheries again or to clean up industrial dump sites such as Love Canal. This approach is based on the implicit assumption that control of emissions and cleanup of natural areas can alleviate the adverse environmental impacts of human economic activity.

As implemented in environmental regulatory practice, this mindset has resulted in a focus on manufacturing activities. All existing major environmental laws in the United States (e.g., the Clean Air Act, the Clean Water Act, the Resource Conservation and Recovery Act, and the Comprehensive Environmental Response, Compensation and Liability Act) deal almost entirely with industrial emissions or the sites of previous industrial emissions or waste disposal. Regulations have identified and, in many cases, mandated specific emission control technologies for such point sources. They have less frequently attempted to deal with geographically dispersed nonpoint sources, such as agricultural runoff. This approach has led to instances of significant short-term reductions in pollution—the Hudson is indeed cleaner than it was 15 years ago. It has also begged the inevitable questions associated with the more fundamental restructuring of technology and economic activity that will undoubtedly be required if a stable long-term global carrying capacity for the human species is to be achieved.

---

[*]A version of this paper was published previously in *Industrial Ecology*. ©1995 Prentice-Hall. Reprinted by permission.

It is now apparent that this first, naive view of the interaction of the global economy with natural environmental systems is simplistic and inadequate. It must be replaced with a more systems-based approach that goes beyond localized phenomena and integrates environment and technology throughout all human economic activity. This nascent, multidisciplinary field is known as industrial ecology and is being implemented in private firms in the manufacturing sector through methodologies and tools developed using design-for-environment (DFE) approaches (Allenby, 1992, 1994a; American Electronics Association, 1993). DFE programs may, in turn, be divided into two categories: generic DFE, which includes things such as "green" accounting systems (Todd, 1994) to improve the environmental performance of the firm as a whole, and specific DFE, which focuses on tools applied to the design of individual manufacturing processes and products (Allenby, 1994b; Glantschnig, 1992).

The relationship between past approaches for evaluating the environmental impacts of human activities and industrial ecology is captured in Table 1. Note the shift in emphasis from specific wastes and materials to products as they are actually used in commerce, and from a geographically and temporally localized view of environmental insults to a regional and global view. This shift recognizes that local insults must be remedied but that the environmental perturbations of real concern relate to the broader issues of human population growth, loss of biodiversity, global climate change, ozone depletion, and depletion of water and arable soil.

It is worth emphasizing that the past (remediation) and present (compliance) approaches are closely linked and generally require similar competencies. Industrial ecology is far broader in its economic and environmental implications and requires very different competencies (e.g., strategic planning). It is different in kind, not just degree, from the mindset behind both the remediation and the compliance approaches to environmental perturbations.

What is the implication of this new philosophy for facilities? For one, industrial ecology requires that facilities of all types be subject to the same scope of evaluation as product or process design. Facilities must be evaluated in terms of the materials with which they are constructed, how they are used (analogous to process technology issues), and how they are refurbished and reused (analogous to product-life extension). As with other DFE efforts, the goal is to design, purchase, or adapt facilities in an environmentally responsible manner that contributes to their competitive advantage. This matrix tool, therefore, should be regarded as only one component of the full DFE set that must be developed as firms begin implementing the principles of industrial ecology.

THE ENVIRONMENTALLY RESPONSIBLE FACILITY

Two aspects of industrial ecology/DFE are critical for the environmentally responsible facility (ERF). The first is the emphasis in any DFE analysis on a

TABLE 1 Evolution of Environmental Regulation

Time Focus	Principal Activity	Focus of Activity	Geographic/Temporal Scale	Endpoints	Key Competencies	Regulatory Approach	Government/Institutional Leaders
Past	Remediation	Waste substances/Sites	Local/Immediate	Reduction of immediate human risk	Toxicology, environmental science	Command and control	U.S. environmentalists
Present/Emphasis on past	Compliance	Emitted substances; emphasis on end-of-pipe control	Point source/Immediate	Reduction of immediate human risk	Toxicology, environmental science, environmental engineering	Command and control; mandated end-of-pipe technologies	Developed countries/Environmentalists
Present/Looking toward future	Industrial ecology/Design for environment	Products and services over life cycle/Industrial and consumer behavior in actual economy and resulting environmental impacts	Regional and global systems at all time scales	Sustainability, including global climate change; loss of biodiversity; degradation of water, soil, and atmospheric resources; ozone depletion	Physical and biological sciences; engineering (especially chemical engineering) and technology; environmental science; business; law; and economics	Product life-cycle regulation (e.g., product take-back); market incentives for environmentally appropriate behavior (e.g., ecolabels; "energy star," international standards; "green procurement")	European Union, especially Netherlands and Germany; industry, especially electronics

systems-based, life-cycle approach. As applied to facilities, this means that both the initial siting decision and the decision to refurbish, sell, or close the facility should take into account the environmental implications of those actions. The second is to realize that this field is in a nascent stage of development. What we present here, therefore, represents an initial effort to define a DFE tool to evaluate ERFs, which we anticipate will be considerably elaborated in the future.

Any methodology that is to be broadly applicable to facilities must be process rather than technology oriented. A fast-food restaurant and a silicon chip manufacturing plant are vastly different in function and technology, yet it is appropriate and necessary to make the same basic evaluations of both. The tool we are proposing here is designed to establish and support a generally applicable assessment process. In practice, however, characteristics specific to the location, purpose, local ecology and demographics, and embedded technology of each facility will come into play in performing the evaluation.

Experience appears to demonstrate that a life-cycle assessment (LCA) of a complex facility is most effective when it is done in modest depth and in a qualitative manner by an industrial ecology specialist. To facilitate such assessments, we have devised a standardized environmentally responsible facility matrix, supported by a checklist to guide assessors in valuing the matrix elements. The matrix scoring system provides a straightforward means of comparing options, and dot charts are recommended as a convenient and visually useful way of calling attention to those design and implementation aspects of the facility whose modification could most dramatically improve the ERF rating.

ERF assessment need not and should not be applied only to manufacturing facilities. Any facility providing products or services—oil refineries, auto body shops, fast-food restaurants, office buildings, and so forth—can benefit from the approach. It would not be unreasonable, in fact, for developers of private housing to use this methodology, if incentives could be created for them to do so.

The ERF Matrix

A suitable ERF assessment system should:

- allow direct comparisons among facilities,
- be usable and reasonably consistent across different assessment teams,
- encompass all stages of facility operations and all relevant environmental concerns, and
- be simple enough to permit relatively quick and inexpensive assessments.

The central feature of the system we recommend is a five-by-five matrix, one dimension of which is environmental concern, the other of which is facility activities (Table 2). The assessor studies the different activities within the facility and their impacts and assigns to each element of the matrix a rating from 0 (high-

TABLE 2   Environmentally Responsible Facility Assessment Matrix

	Environmental Concern[a]				
	Ecological Impacts	Energy Use	Solid Residues	Liquid Residues	Gaseous Residues
Site selection, development, and infrastructure	1,1	1,2	1,3	1,4	1,5
Principal business activity—products	2,1	2,2	2,3	2,4	2,5
Principal business activity—processes	3,1	3,2	3,3	3,4	3,5
Facility operations	4,1	4,2	4,3	4,4	4,5
Facility refurbishment, closure, or transfer	5,1	5,2	5,3	5,4	5,5

[a]The number in each cell corresponds to the relevant question set for that cell, as outlined in the Appendix.

est impact, a very negative evaluation) to 4 (lowest impact, an exemplary evaluation). The ERF rating is the sum of the matrix element values. Because there are 25 matrix elements, the best facility rating is 100.

In arriving at an individual matrix element assessment, or in offering advice to managers seeking to improve the rating of a particular matrix, the assessor uses detailed checklists and special evaluation techniques. Many checklist items will be common to all facilities, whereas others will be specific to the activity of the particular facility. An illustrative ERF checklist system for a generic manufacturing facility appears as an appendix to this paper.

The assignment of discrete values from zero to 4 for each matrix element assumes that the DFE implications of each element are equally important. The utility of the assessment might be increased by applying weighting factors to the matrix elements, although this may also increase the complexity of the task. For example, if global warming impacts of a facility's operations were judged to outweigh the localized impacts of liquid residues, weighting of the "energy use" column could be increased and that of the "liquid residue" column correspondingly decreased. When comparing facilities or assessments with one another, of course, identical weighting factors must be used.

This system is deliberately semiquantitative to respond to the conundrum that has often bedeviled attempts to develop workable DFE/LCA tools. On the one hand, it is extremely difficult—many professionals would say impossible—to quantify the impacts of even those environmental releases and effects that can be inventoried. For example, how should one quantitatively evaluate the trade-offs between using a substance with a highly uncertain potential for human carcinogenicity and one tied to possible loss of biodiversity? (What is the value of a species, and is it ethical even to pose such a question?) On the other hand, quantitative systems are a prerequisite for diffusion of DFE methodologies and concepts throughout industry, especially if modifications to business planning and design processes are desired. The ERF matrix system thus explicitly relies on the professional judgment of industrial ecologists, while allowing for standardization of dimensions through common checklists as the state of the art advances and experience is gained. The system provides an easily used management and operational tool, but it does not pretend to greater certainty than the underlying data justify.

Matrix Structure

The columns of the matrix correspond to the five major classes of environmental concern: ecological impacts, energy use, solid residues, liquid residues, and gaseous residues. Although other categories could no doubt be suggested, these are readily understood and reasonably comprehensive, in keeping with the practical intent of the system. Both local ecological impacts and (if applicable) loss of biodiversity could be included in the first column, for example.

The rows correspond to the life cycle of a generic facility (modified slightly to fit the manufacturing example we are using). As these are less intuitive (even environmental professionals are not yet familiar with the concept of the life cycle of a facility), a more detailed description of each life-cycle stage is appropriate.

Site Selection, Development, and Infrastructure

A significant factor in evaluating the degree of a facility's environmental responsibility is the site selected and the way in which the site is developed. If the facility is an extraction or materials-processing operation (e.g., oil refining or ore smelting), the location will generally be constrained by the need to be proximate to the resource. A manufacturing facility usually requires access to good transportation and a suitable workforce but otherwise may be unconstrained. Service facilities usually must be located near customers. Office buildings may be located virtually anywhere, so long as it is reasonably possible for employees to commute. Housing developments must be located where people want to live. In all cases, it might be possible to refurbish or add new operations to existing facilities, avoiding many of the regulatory difficulties and environmental impacts of establishing a "greenfield" facility site.

Manufacturing plants have traditionally been located in or near urban areas. Such siting has the advantages of drawing on a geographically concentrated workforce and of using existing transportation and utility infrastructures. One problem with urban sites in some countries is that there may be laws that force purchasers of property formerly used for commercial or industrial purposes to assume liability for any environmental damage caused by the previous owner or owners. The result has been that urban industrial areas, which from an industrial ecology standpoint are in many ways ideal locations for industrial facilities, have been virtually impossible to use. The governmental and legal systems need to devise a means around this difficulty. (Environmental liability difficulties in urban areas are sometimes secondary to such factors as crime, congestion, and high taxes [Boyd and Macauley, 1994]).

For facilities of any kind built on land not previously used for industrial or commercial purposes, one can anticipate that there will be ecological impacts on regional biodiversity as well as added air emissions (from construction and use of new transportation and utility infrastructures) and water emissions (from sanitary facilities and manufacturing activities). These effects can be minimized by using as much as possible existing infrastructures and developing the site by leaving the maximum area in its natural form. Nonetheless, given the current overstock of commercial buildings and facilities in many countries, such "greenfield" choices are hard to justify from an industrial ecology perspective.

Evaluation of existing infrastructure also requires consideration, and possibly redesign, of other local operations. Within each facility, for example, it is sometimes possible to use a residue stream from one process as a feed stream for another, to use excess heat from one process to provide heat for another, and so on. Such actions constitute steps toward a facility ecosystem. Chemical manufacturing plants, in particular, have made good progress along these lines.

Opportunities also exist to establish portions of industrial ecosystems when facilities owned by different parent companies agree to share residual products or residual energy. Such an approach is encouraged by geographical proximity. For example, the AT&T manufacturing plant in Columbus, Ohio, is about 1 km from a solid-waste landfill that emits methane gas, a by-product of the biodegradation of landfilled material. AT&T purchases the gas from the landfill and pipes it to its plant boiler, where the gas furnishes up to 25 percent of the necessary energy for manufacturing. At the same time, emissions of methane into the air, a greenhouse gas, are reduced.

More complex arrangements are possible, especially if planning is done before facilities are built. These involve establishing close relationships with suppliers, customers, and neighboring industries, and working with those partners to close materials cycles. In the same way that close relationships promote just-in-time delivery of supplies and components, so, too, can those relationships help corporations implement environmentally responsible manufacturing.

An outstanding and still unique example of the partnership approach exists in Kalundborg, Denmark, where 10 years of effort have culminated in the interactive network shown in Figure 1 (Graedel and Allenby, 1995; Terp, 1991). Four main participants are involved: the Asnaesverket Power Company, a Novo Nordisk pharmaceutical plant, a Gyproc facility for producing wallboard, and a Statoil refinery. Steam, gas, cooling water, and gypsum are exchanged among the participants, and some heat also is used for fish farming and residential green-

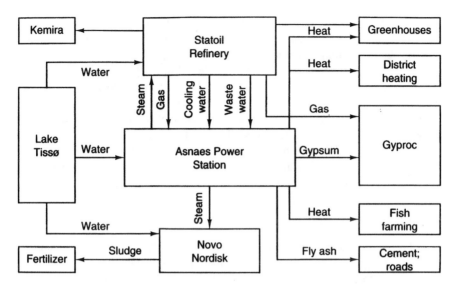

FIGURE 1   Industrial ecosystem at Kalundborg, Denmark.

house heating. Residual products not usable in the immediate vicinity, such as fly ash and sludge, are sold for use elsewhere. None of the arrangements were required by law; rather, all were negotiated independently for reasons of better materials prices or avoidance of materials disposal costs. It is probably accurate to refer to this cooperative project as an early model of an industrial ecosystem. The Kalundborg experience provides a model for industrial ecology at the ecopark level, especially where industrial activities occur in close proximity to one another.

Principal Business Activity—Products

Clearly, any facility that generates products or activities that are environmentally inappropriate should not be considered an ERF. Thus, for example, an otherwise environmentally appropriate manufacturing facility that makes widgets whose design does not permit them to be recycled cannot under most circumstances be considered an ERF, regardless of how well designed it is in other aspects. Evaluating this aspect of the ERF will require analysis of the output of the facility, whatever that may be. If the output is a product, the environmentally responsible product matrix system can be used (Graedel and Allenby, 1995).

Principal Business Activity—Processes

As with products, it is apparent that, for a facility to be environmentally responsible, its internal processes must also be environmentally responsible. For manufacturing facilities, for example, this means that emissions of residues from all processes should be evaluated and the amount of residue converted to waste minimized, the use of toxic materials in processes should be minimized, and the appropriate emission controls should be installed. Similarly, for an office building from which services are provided, the amount of paper used in the processes underlying the service should be minimized, and the use of recycled paper in all operating processes should be maximized. The use of recycled paper in customer billing, for example, would help make the facility within which the billing operation is housed an ERF, all other things being equal. Evaluation of this aspect of a facility's processes can be accomplished using the environmentally responsible process matrix system (Graedel and Allenby, 1995).

Facility Operations

Facility operations can involve a host of disparate activities. For example, the impact of any facility on the environment is heavily weighted by transportation. As with many other aspects of industrial ecology, trade-offs are involved. For example, just-in-time delivery of components and modules has been hailed as cost effective and efficient. Nonetheless, it has been estimated that the largest contributor to the Tokyo smog problem is trucks making just-in-time deliveries. The corporations delivering and those receiving the components and modules bear some degree of responsibility for these emissions. It is sometimes possible to reduce transport demands by improved scheduling and coordination, perhaps in concert with nearby industrial partners. And there may be options that encourage ride sharing, telecommuting, and other activities that reduce overall emissions from employee vehicles.

Material entering or leaving a facility also offers opportunities for useful action. To the extent that the material is related to products, it is captured by the product DFE assessments. Facilities receive and disperse much nonproduct material, however, including food for employee cafeterias, office supplies, restroom supplies, maintenance items such as lubricants, fertilizer, pesticides, herbicides, and road salt. Frequently, materials and other inputs to a facility are "overpackaged," resulting in substantial unnecessary waste generation. Packaging recycling programs and pressure on suppliers to use environmentally conscious packaging can cut such material consumption significantly. An ERF should have a structured program to evaluate each incoming and outgoing materials stream and to tailor it and its packaging in environmentally responsible directions.

The use of energy by a facility requires careful scrutiny as well, because opportunities for improvement are always present. An example is industrial lighting systems, whose energy needs account for between 5 and 10 percent of air pollution from power plant emissions (in the form of CO_2, SO_2, heavy metals, and particulates). As with many environmentally related business expenditures, lighting costs are often lumped in with overhead and therefore are not known precisely. The use of modern technology has the potential to decrease electrical expenditures for lighting by 50 percent or more. To promote these changes, the U.S. Environmental Protection Agency has initiated the Green Lights program, which encourages the use of high-efficiency fluorescent ballasts and lamps, automatic shut-off of lights when not in use by means of occupancy sensors, and mirrorlike reflectors in existing fluorescent systems (Hoffman, 1992). Corporations agreeing to participate in this voluntary program commit to surveying their lighting and upgrading their systems in ways that reduce pollution, improve the quality of lighting, and still allow for profit goals to be met. Several states and several hundred corporations have agreed to participate.

A routine part of facility operation is the care of the land surrounding buildings or other structures. It is increasingly common to allow that land to serve as a habitat for local flora and fauna. (See, for example, Skinner, 1994.) Such use is good for the environment, public relations, and employee morale, and the elimination of the need for regular maintenance often results in cost savings as well.

Facility Refurbishment, Closure, or Transfer

Just as environmentally responsible products are being designed increasingly for "product-life extension," so too should ERFs be designed for easy upgrading. Buildings contain substantial amounts of material with significant embedded energy, and the environmental disruption (particularly in the local area) involved in constructing buildings and related infrastructure is significant. In the United States, construction accounts for the largest use of material by far. In 1990, for example, some 2.53 billion metric tons of materials were consumed, of which about 70 percent, or 1.75 billion metric tons, was construction materials (Bureau of Mines, 1993). Clearly, an ERF must be designed to be easily refurbished for new uses, to be transferred to new owners and operators with a minimum of alteration, and, if it must be closed, to permit recovery for reuse and recycling of materials, fixtures, and other components. To some extent, the first two requirements are taken into consideration today, but, in general, the latter is almost never recognized as an important design feature of new facilities.

Construction of Dot Charts

After the overall rating for a facility is determined, the use of a dot chart will provide a succinct display of the results and facilitate the identification of issues that should be given special attention. Such a plot is shown in Figure 2, constructed using illustrative data. Outliers can be readily identified. In the example, the greatest opportunity for improvement lies at points 2, 5; 4, 2; and 5, 3. Alternative facility locations or different designs for environmental preferability can be easily compared using dot charts.

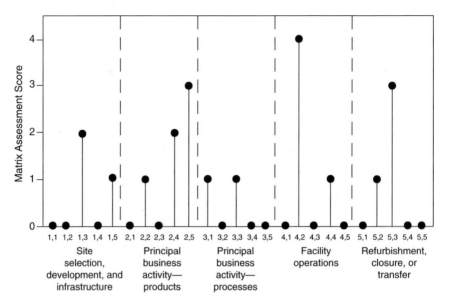

FIGURE 2   Dot chart representation of hypothetical ERF matrix results.

CONCLUSION

Despite rhetoric to the contrary and years of practice, a true systems-based, life-cycle approach to assessing facilities' environmental impacts virtually does not exist. Emissions have been targeted for regulation for years, but the concept that the facility itself should be designed to be environmentally "friendly" over its lifetime has never been explored, in spite of the enormous environmental impacts of construction and development. This in itself is a significant indictment of the current fragmented, ad hoc system of environmental management and regulation, and a clear demonstration of the need to move toward policies and practices based on industrial ecology principles. The ERF matrix system begins the process of thinking about facilities from the life-cycle perspective, developing analytical tools and, as experience is gained, metrics to support the LCA of environmental impacts. Nonetheless, it must also be remembered that facilities are only part of the economic stream from which environmental impacts flow and that the matrix system is only one of many analytical tools that will be required.

Unlike some LCA activities, overall LCA as presented here is less quantifiable and less thorough. It is also more practical. A survey of the modest depth that we advocate, performed by an objective professional, will succeed—for a relatively small investment of time and money—in identifying perhaps 80 or 90 percent of useful facility-related DFE actions that could be taken. It is far better to conduct a number of modest LCAs than to conduct one or two in great depth.

Furthermore, it is critical to recognize that using these practical tools, even if they are open to criticism by purists, represents a substantial advance over any practices currently in place.

The key ingredient in a successful LCA is the expert who performs it. This person, whether from inside or outside the corporation, must be experienced and knowledgeable about the types of products, processes, and facilities being reviewed. This is a lot to ask, but no more than if the same person were to perform a classical LCA with the same goals.

Improvement analysis—and the actual implementation of the identified improvements—is the ultimate goal of all industrial ecology activities. As with most ecological situations, however, the actions taken will reflect a variety of trade-offs. One should not enter into a life-cycle analysis of a facility with the idea that all possible actions can be accomplished. Rather, the process helps identify elements of facility design or operations that might be modified; facility managers must decide which are practical to implement. The result will in each case be a facility that is much more environmentally sustainable than if nothing had been done.

NOTE

1. As part of AT&T's ongoing effort to implement industrial ecology through development of DFE methodologies, the company is creating a family of matrices, including ones dealing with environmentally responsible products and processes as well as facilities. The aim is to provide straightforward and easily used capability to perform life-cycle DFE assessments across major activities and operations of private firms.

APPENDIX

THE ENVIRONMENTALLY RESPONSIBLE FACILITY MATRIX

In this appendix, a sample of possible items appropriate to each of the matrix elements for the environmentally responsible facility matrix tool is presented. It is anticipated that different types of facilities will require different checklists and evaluations, so this appendix is presented as an example rather than as a universal formula.

Facility Matrix Element 1,1
Facility Activity: Site Selection, Development, and Infrastructure
Environmental Concern: Ecological Impacts

• Has the proposed site previously been used for similar activities? If not, have any such sites been surveyed for availability?
• Is necessary development activity, if any, being planned to avoid disruption of existing biological communities?
• Are the biota of the site compatible with all planned process emissions, including possible emissions that exceed allowable levels?
• Has the site been chosen to minimize the need for new on-site infrastructure (buildings, roads, etc.)?
• If new infrastructure must be created, are plans in place to minimize any resulting impacts on biota?
• Have provisions been made for orderly growth of infrastructure as facility operations expand, in order to avoid unnecessary health or environmental impacts?

Facility Matrix Element 1,2
Facility Activity: Site Selection, Development, and Infrastructure
Environmental Concern: Energy Use

• Is the site such that the facility can be made operational with only minimal energy expenditures?
• Has the site been selected to avoid any energy emission impacts on existing biota?
• Does the site allow delivery and installation of construction or renovation materials with minimal use of energy?
• Does existing energy infrastructure (gas pipelines, electric power cable) reduce or eliminate the need to build new systems?

• Is it possible to use heat residues from within the plant or from nearby facilities owned by others to provide heat or power, or to cogenerate for them?
• Is it possible to use gaseous residues from within the plant or from nearby facilities owned by others to provide heat or power, or to cogenerate for them?

Facility Matrix Element 1,3
Facility Activity: Site Selection, Development, and Infrastructure
Environmental Concern: Solid Residues

• Is the site such that the facility can be made operational with only minimal production of solid residues?
• Have plans been made to ensure that any solid residues generated in the process of developing the site are managed to minimize their impacts on biota and human health?
• If any solid residues generated in the process of developing the site are hazardous or toxic to the biota or humans, have plans been made to minimize releases and exposures?
• Is it possible to use as feedstocks solid residues from nearby facilities owned by others?
• Is it possible to use solid residues from the proposed facility as feedstocks for nearby facilities owned by others?
• Can the transport and disposal of solid residues be shared with nearby facilities owned by others?

Facility Matrix Element 1,4
Facility Activity: Site Selection, Development, and Infrastructure
Environmental Concern: Liquid Residues

• Is the site such that the facility can be made operational with only minimal production of liquid residues?
• Have plans been made to ensure that any liquid residues generated in the process of developing the site are managed to minimize their impacts on biota and human health?
• If any liquid residues generated in the process of developing the site are hazardous or toxic to biota or humans, have plans been made to minimize releases and exposures?
• Is it possible to use as feedstocks liquid residues from nearby facilities owned by others?
• Is it possible to use liquid residues from the proposed facility as feedstocks for nearby facilities owned by others?
• Can the transport and disposal of liquid residues be shared with nearby facilities owned by others?

Facility Matrix Element 1,5
Facility Activity: Site Selection, Development, and Infrastructure
Environmental Concern: Gaseous Residues

• Is the site such that the facility can be made operational with only minimal production of gaseous residues?

• Have plans been made to ensure that any gaseous residues generated in the process of developing the site are managed to minimize their impacts on biota and human health?

• If any gaseous residues generated in the process of developing the site are hazardous or toxic to biota or humans, have plans been made to minimize releases and exposures?

• Is it possible to use gaseous residues from the proposed facility to provide heat or power for nearby facilities owned by others?

• Is it possible to use gaseous residues from the proposed facility to provide process or product feedstocks for nearby facilities owned by others?

• Is it possible to share employee transportation infrastructure with nearby facilities owned by others to minimize air pollution by private vehicles?

Facility Matrix Element 2,1
Facility Activity: Principal Business Activity—Products
Environmental Concern: Ecological Impacts

• If the activity of this facility involves extraction of virgin materials, is the extraction planned so as to minimize ecological impacts, and have restoration plans been made and funding assured, as appropriate?

• Do all outputs from the site, including residue streams, have high ratings as environmentally responsible products?

• Are products designed to use recycled materials?

• Have all outputs from the site been dematerialized to the fullest extent possible?

Facility Matrix Element 2,2
Facility Activity: Principal Business Activity—Products
Environmental Concern: Energy Use

• Are products designed to require minimal consumption of energy in manufacture?

• Are products designed to require minimal consumption of energy in use?

• Are products designed to require minimal consumption of energy in recycling or disposal?

Facility Matrix Element 2,3
Facility Activity: Principal Business Activity—Products
Environmental Concern: Solid Residues

• Are products designed to generate minimal and nontoxic solid residues during manufacture?
• Are products designed to generate minimal and nontoxic solid residues during use?
• Are products designed to generate minimal and nontoxic solid residues when recycled or disposed of?

Facility Matrix Element 2,4
Facility Activity: Principal Business Activity—Products
Environmental Concern: Liquid Residues

• Are products designed to generate minimal and nontoxic liquid residues during manufacture?
• Are products designed to generate minimal and nontoxic liquid residues during use?
• Are products designed to generate minimal and nontoxic liquid residues when recycled or disposed of?

Facility Matrix Element 2,5
Facility Activity: Principal Business Activity—Products
Environmental Concern: Gaseous Residues

• Are products designed to generate minimal and nontoxic gaseous residues during manufacture?
• Are products designed to generate minimal and nontoxic gaseous residues during use?
• Are products designed to generate minimal and nontoxic gaseous residues when recycled or disposed of?

Facility Matrix Element 3,1
Facility Activity: Principal Business Activity—Processes
Environmental Concern: Ecological Impacts

• Have all process materials been optimized from a design-for-environment standpoint?
• Have processes been dematerialized (evaluated to ensure that they have minimum resource requirements and that no unnecessary steps are required)?
• Do processes generate waste heat or emit residues that have the potential

to harm local or regional biological communities, and, if so, have capture and reuse of these resources been explored?

Facility Matrix Element 3,2
Facility Activity: Principal Business Activity—Processes
Environmental Concern: Energy Use

• Have all process materials been evaluated to ensure that they use as little energy as possible?
• Are processes monitored and maintained on a regular basis to ensure that they retain their energy efficiency as designed?
• Do process equipment specifications and standards require the use of energy-efficient components and subassemblies?

Facility Matrix Element 3,3
Facility Activity: Principal Business Activity—Processes
Environmental Concern: Solid Residues

• Are processes designed to generate minimal and nontoxic solid residues?
• Where solid materials are used as process inputs, have attempts been made to use recycled materials?
• Are processes designed to produce usable by-products, rather than by-products suitable only for disposal?

Facility Matrix Element 3,4
Facility Activity: Principal Business Activity—Processes
Environmental Concern: Liquid Residues

• Are processes designed to generate minimal and nontoxic liquid residues?
• Where liquid materials are used as process inputs, have attempts been made to use recycled materials?
• Are pumps, valves, and pipes inspected regularly to minimize leaks?

Facility Matrix Element 3,5
Facility Activity: Principal Business Activity—Processes
Environmental Concern: Gaseous Residues

• Are processes designed to generate minimal and nontoxic gaseous residues?
• Are processes designed to avoid the production and release of odorants?
• If volatile organic compounds are utilized in any processes, are they selected so that any releases will have minimal photochemical smog impact?
• If greenhouse gases, particulates, or nitrogen or sulfur oxides are generated, are they captured and have less environmentally harmful options been evaluated?

Facility Matrix Element 4,1
Facility Activity: Facility Operations
Environmental Concern: Ecological Impacts

• Has the maximum possible portion of the facility been returned to, or allowed to remain in, its natural state?
• Is the use of pesticides, herbicides, fertilizers, or any other chemical treatments on the property minimized?
• Is noise pollution from the site minimized?

Facility Matrix Element 4,2
Facility Activity: Facility Operations
Environmental Concern: Energy Use

• Is the energy needed for heating, ventilating, and cooling the facility minimized?
• Is the energy needed for lighting the facility minimized?
• Is energy efficiency a consideration when buying or leasing facility equipment such as copiers, computers, and fan motors?
• Have maintenance programs been designed and implemented to maintain peak energy efficiency of all systems?
• Has the possibility of on-site generation of energy in environmentally preferable ways been explored?

Facility Matrix Element 4,3
Facility Activity: Facility Operations
Environmental Concern: Solid Residues

• Is the facility designed to minimize the comingling of solid-waste streams?
• Are solid residues from facility operations reused or recycled to the maximum extent possible?
• Are unusable solid residues from facility operations (including food service) disposed of in an environmentally responsible manner and as close to the facility as possible?

Facility Matrix Element 4,4
Facility Activity: Facility Operations
Environmental Concern: Liquid Residues

• Is the facility designed to minimize the comingling of liquid-waste streams?
• Are liquid treatment plants monitored to ensure that they operate at peak efficiency?

• Have liquid residue waste streams been reviewed to determine if they can be redesigned to be commercially valuable?

• Are unusable liquid residues from facility operations disposed of in an environmentally responsible manner?

Facility Matrix Element 4,5
Facility Activity: Facility Operations
Environmental Concern: Gaseous Residues

• Is operations-related transportation to and from the facility minimized?

• Are furnaces, incinerators, and other combustion processes and their related air pollution control devices monitored to ensure they are operating at peak efficiency?

• Is employee commuting minimized by job sharing, telecommuting, and similar programs?

Facility Matrix Element 5,1
Facility Activity: Facility Refurbishment, Closure, or Transfer
Environmental Concern: Ecological Impacts

• Will activities necessary to refurbish, close, or transfer the facility to alternate uses cause any ecological impacts and, if so, has planning been done to minimize such impacts?

• When refurbishment, closure, or transfer activities are undertaken, can the materials used and any surplus materials be recycled with a minimum of ecological impact?

• Has a "facility-life extension" review been undertaken to optimize the life and service of the existing facility, therefore minimizing the need to construct new facilities with their attendant environmental impacts?

Facility Matrix Element 5,2
Facility Activity: Facility Refurbishment, Closure, or Transfer
Environmental Concern: Energy Use

• Can the facility be closed or transferred with a minimum expenditure of energy (including any necessary site cleanup and decontamination)?

• Can the facility be modernized and converted to other uses easily?

• When the facility is refurbished, closed, or transferred, has it been designed and are plans in place to recapture as much of the embedded energy as possible?

Facility Matrix Element 5,3
Facility Activity: Facility Refurbishment, Closure, or Transfer
Environmental Concern: Solid Residues

• Can the facility be refurbished, closed, or transferred with minimal generation of solid residues, including those generated by site cleanup and decontamination?

• At closure, can the materials in the facility, including all structural material and remaining capital stock, be reused or recycled with minimal generation of solid residues?

• Have plans been made to minimize the toxicity of and exposures to any solid residues resulting from cleanup and decontamination of the facility and its environs upon refurbishment, transfer, or closure?

Facility Matrix Element 5,4
Facility Activity: Facility Refurbishment, Closure, or Transfer
Environmental Concern: Liquid Residues

• Can the facility be refurbished, closed, or transferred with minimal generation of liquid residues, including those generated by site cleanup and decontamination?

• At closure, can the materials in the facility, including all structural material and remaining capital stock, be reused or recycled with a minimal generation of liquid residues?

• Have plans been made to minimize the toxicity of and exposures to any liquid residues resulting from cleanup and decontamination of the facility and its environs upon refurbishment, transfer, or closure?

Facility Matrix Element 5,5
Facility Activity: Facility Refurbishment, Closure, or Transfer
Environmental Concern: Gaseous Residues

• Can the facility be refurbished, closed, or transferred with minimal generation of gaseous residues, including those generated by site cleanup and decontamination?

• At closure, can the materials in the facility, including all structural material and remaining capital stock, be reused or recycled with minimal generation of gaseous residues?

• Have plans been made to minimize the toxicity of and exposures to any gaseous residues resulting from cleanup and decontamination of the facility and its environs upon refurbishment, transfer, or closure?

REFERENCES

Allenby, B. R. 1992. Design for Environment: Implementing Industrial Ecology. Ph.D. dissertation, Rutgers University, New Brunswick, N.J.

Allenby, B. R. 1994a. Industrial ecology gets down to earth. IEEE Circuits and Devices 10(1):24–28.

Allenby, B. R. 1994b. Integrating environment and technology: Design for environment. Pp. 137–148 in The Greening of Industrial Ecosystems, B. R. Allenby and D. J. Richards, eds. Washington, D.C.: National Academy Press.

American Electronics Association. 1993. The Hows and Whys of Design for the Environment—A Primer for Members of the American Electronics Association. Washington, D.C.: American Electronics Association.

Boyd, J., and M. K. Macauley. 1994. The impact of environmental liability on industrial real estate development. Resources (Winter):19–23.

Bureau of Mines. 1993. Materials and environment: Where do we stand? Minerals Today (April):6–13.

Glantschnig, W. 1992. Design for environment (DFE): A systematic approach to green design in a concurrent engineering environment. In Proceedings of the First International Congress on Environmentally Conscious Design and Manufacturing, May 4–5, 1992, Boston, Mass.

Graedel, T. E., and B. R. Allenby. 1995. Industrial Ecology. Englewood Cliffs, N.J.: Prentice-Hall.

Hoffman, J. S. 1992. Pollution prevention as a market-enhancing strategy: A storehouse of economical and environmental opportunities. Proceedings of the National Academy of Sciences 89:832–834.

Skinner, J. P. 1994. Chemical companies go for greener pastures. Today's Chemist at Work (March):40–48.

Terp, E. 1991. Industrial symboise i Kalundborg. In the corporate report of the Asnaesverket Electric Power Co., Kalundborg, Denmark.

Todd, R. 1994. Zero-loss environmental accounting systems. Pp. 191–200 in The Greening of Industrial Ecosystems, B. R. Allenby and D. J. Richards, eds. Washington, D.C.: National Academy Press.

Accounting Methods

Measures of Environmental Performance and
Ecosystem Condition. 1999. Pp. 65–69.
Washington, DC: National Academy Press.

Measuring Pollution-Prevention Performance

THOMAS W. ZOSEL

The demand for effective measures of pollution-prevention performance has increased with industry's concern about environmental issues. Many corporations spend significant portions of their capital and operating budgets to address environmental issues, and corporate managers need ways to measure the results of these efforts. It is the job of the environmental staff within a corporation to determine what measurements need to be made and reported to top management.

Many others are also interested in the environmental performance of a company or facility. For example, the communities in which plants are located may be extremely concerned about what is released into the environment. Customers, too, have an interest and potentially a legal right to know what materials are in products or are used during their manufacture. Of particular legal interest today is whether chlorofluorocarbons are used in a product's manufacture. Finally, environmental agencies have a distinct interest in tracking environmental performance. However, because agencies' interests are manifested in legal and regulatory requirements, companies have little choice in the type of measurements they make. Permits, plant operations, and—because the criminal sanctions in many laws are enforced—employees' personal freedom may depend on making and reporting the required measurements. Given the number of stakeholders, no single environmental metric or measurement system is likely to meet everyone's needs.

TOXIC RELEASE INVENTORY

Perhaps the data that have received the most publicity in the past few years are the Superfund Amendments and Reauthorization Act of 1986 (SARA) Toxic Release Inventory (TRI) reports. These reports list quantities of selected toxic chemicals that have been released by a facility into the environment. Such data provide some information about environmental performance. However, SARA TRI numbers do not take into account the fact that quantity is not the only measurement of risk. A small quantity of a potent carcinogen might pose a significantly greater health risk than a large quantity of a mild irritant.

Even with that inherent limitation, TRI reports have become the focal point for environmental groups, local communities, and the media, primarily because the numbers are easy to understand. The media find the numbers of particular value. Current and historical releases, expressed as pounds per year of a particular chemical, can easily be compared. In addition, because all major manufacturing facilities are required to file these reports annually, facilities can be compared with one another.

Such comparisons are used by environmental organizations to push plants with high releases to meet the lower emissions levels achieved by comparable facilities. There have even been suggestions that performance comparisons be required by law. Today, in air regulation, new sources of emissions in nonattainment areas are required to meet lowest-achievable rate standards—the lowest emission rate that is achieved in practice or is required by any law or regulation in that state.

Similarly, new plants could be required to meet a lowest-achievable release rate. This would be the release rate to all media achieved by a similar manufacturing facility with the best performance. This type of regulation could be extremely burdensome and raises some very disturbing issues regarding proprietary information. The basic concern is that the plants that perform best may be achieving these exceptional results through the use of proprietary technology. Could this type of legislation require these companies to share their superior yet proprietary technology with a competitor? Even if a licensing fee were offered, the mandatory sharing of such technology would create a significant monetary and competitive loss for the company that developed it.

Although there are many ways to use the SARA TRI information, data requirements and report formats are mandated by regulatory agencies. These data may not satisfy a company's needs. Thus, companies may also want to consider developing other metrics that are more useful for internal measurement of environmental performance.

POLLUTION PREVENTION AT 3M

At 3M, we have quantitatively measured environmental performance since 1975. The system in use quantifies the pollution that has been prevented under 3M's Pollution Prevention Pays (3P) program and the resulting monetary savings. Within this system, 3M defines pollution prevention as source reduction and environmentally sound reuse and recycling. Although this metric does not indicate total environmental performance, it does address an extremely important issue for top management. What it tracks is the amount of pollution that has been prevented through cost-effective projects. Since 1975, 3M pollution-prevention projects have stopped roughly 700,000 tons of pollutants from entering the environment and saved the company over $750 million.

The 3P program emphasizes to all employees that they can take actions both to reduce the actual volume of pollution being generated and to increase the monetary savings that result from these actions. However, this metric did not tie reductions in pollution to specific production activities, and it did not include reductions that were achieved but not reported.

In the late 1980s, 3M began looking at ways that it could measure and report waste generation and waste reduction that would better fit into a total quality management (TQM) program. The system needed to be simple, accurate, and reproducible. It needed to be indexed to production so that the waste was viewed in relationship to total plant output. Also, the system needed to measure the reduction in waste resulting from pollution-prevention efforts. Measurements needed to be made before the waste was treated, controlled, or disposed of. The system had to allow the establishment of goals at both the corporate and the division levels. And, perhaps most important, the system needed to motivate employees.

After considerable discussion and trials of pilot projects that used different measurement schemes, 3M implemented such a system in 1990. The system classifies all outputs from a production facility into one of three categories: product, or the intended output from the manufacturing facility; by-product, or residuals that are productively used through some form of recycling or reuse; and waste, or material that is subjected to waste treatment, pollution control, or is directly released into the environment. Together, these items represent the total output from the manufacturing facility. The metric reported is the waste ratio:

waste ratio = waste/(waste + by-product + product) = waste/total output.

This metric is, in a manner of speaking, a measurement of manufacturing efficiency. If the waste ratio is zero, then all raw materials are being productively used. This does not mean that the plant's processes are 100 percent efficient or that all raw materials are being converted to product. For the majority of operations, the latter is impossible.

At 3M, the waste ratio is reported by division, not merely by plant. Manufacturing employees can make significant contributions to reducing waste, but the participation of many other workers is necessary to fully implement waste-reduction and pollution-prevention efforts. Through research and development, company scientists develop new products and processes that generate less waste. 3M engineers design more-efficient equipment that will accept recycled or re-used materials. Finally, the company's sales force preferentially sells products that generate the least waste. In short, waste reduction and pollution prevention are everyone's job. If the objective is to change the corporate culture and the way that companies view environmental waste, then each and every employee must play an integral part in the program.

The objective of the 3M waste-measurement system is to obtain a single number for each division. Existing databases contain extensive information that, if properly integrated, can be used to calculate the amounts of waste generated. Taking advantage of that information, 3M divides waste measurement into five easily measured categories: chemical waste, trash, organic waste (in air and water), particulate waste (in air), and water waste (excluding the water itself). (We do not need to measure each specific waste stream, but we do need to measure all of the wastes generated by a division or facility.)

Chemical Waste

Chemical waste is defined as all the material included on a Resource Conservation and Recovery Act manifest. Determining the quantity generated is as simple as tapping into the database that contains the manifest information. If this information is not in a database, the quantities can be calculated by hand. If manifest information is not available, the facility and the company's management have a much greater problem than measuring waste.

Trash

Most major landfills in the United States weigh the amount of trash sent to their facilities, normally with a truck scale. Therefore, it is generally fairly easy to make arrangements with the landfill operator to obtain this information. If the local landfill does not weigh the trucks, it is relatively simple and inexpensive to find a local truck scale that can make the necessary measurements.

Organic Waste

Measuring organic wastes can be difficult, especially for releases into the air from fugitive sources. The problem can be considerably simplified by taking a materials-balance perspective. Organic waste is the amount of organic material brought into the plant, minus the amount that goes out as product, is shipped off site as chemical waste or for recycling, and is consumed or transformed in a chemical reaction. Each of these individual items is relatively easy to calculate. Purchasing records should show the amount of total volatile organics brought into the plant. The amount shipped as chemical waste should be available from the manifests. The amount shipped for recycling should be available from shipping records. The amount consumed in chemical reactions should be available from production- or process-engineering yield data. The remainder is the volatile organic material that is waste before it is subjected to treatment or pollution control. This information may not indicate the final disposition of the waste, but for the purposes of this metric, that kind of detail is not necessary.

Particulate Waste

The majority of particulate waste will actually be measured in the other waste categories. Material that is collected from dry control systems, such as bag houses, will be either in the chemical-waste measurement if the material is hazardous or in the trash measurement if the material in nonhazardous. If a wet control system is used, the material will end up in the water-waste measurement. With this in mind, the actual particulate waste that is in an air stream is very small and, in most cases, can be ignored. In those few cases where it is relevant, an actual measurement system may be needed.

Water Waste

Because the objective is to obtain a measurement of the total waste, not of the individual components, the amount of waste in the water can be determined by a simple analysis of total dissolved solids. This also takes into consideration that any organics in the water have been previously accounted for in the organics categories.

PROGRESS IN WASTE REDUCTION

The metric and measurement system was implemented throughout 3M in 1990. 3M management established a goal of reducing waste by 35 percent by 1995. The company's Challenge 95 asked each division to accomplish a 7 percent per year reduction in the rate of waste generation during each of the next 5 years.

By the end of 1995, 3M achieved 30 percent waste reduction compared with 1990 levels in the United States and a 32.5 percent reduction worldwide. The waste-ratio measurement system and the establishment of waste-reduction goals have clearly motivated 3M employees to reduce the generation of waste. The company's new goal is to reduce waste production to 50 percent of the 1990 level by 2000.

SUMMARY

The metrics required by laws and regulations can be useful indicators of environmental performance, but they are not always suitable for internal corporate purposes. Using existing databases, 3M has developed a waste metric that conforms to a TQM structure, is simple yet accurate, is a good employee motivator, can be utilized to establish goals, and, after 5 years of experience, has proved effective in accomplishing waste reduction and pollution prevention.

Measures of Environmental Performance and
Ecosystem Condition. 1999. Pp. 70–88.
Washington, DC: National Academy Press.

Accounting for Natural Resources in Income and Productivity Measurements

ROBERT C. REPETTO, PAUL FAETH,
AND JOHN WESTRA

Sustainable development has been defined variously as (a) living on nature's income instead of depleting its capital, (b) meeting the needs of today's population without compromising the ability of future generations to meet theirs, and (c) managing natural, human, and financial assets to increase human health and well-being over the long term. By whatever definition, moving toward sustainable development is clearly in the vital interest of societies everywhere.

Trouble arises when the indices by which we try to measure improvements in living standards ignore the loss of natural resources and the services that they provide. Policy makers, who inevitably rely on these flawed measures of economic development, can get very misleading signals, leading to temporary improvements in consumption that are "purchased" by permanent losses in wealth and productive capacity.

The fundamental definition of income encompasses the notion of sustainability. In accounting and economic textbooks, income is defined as the maximum amount that can be consumed in a given period without reducing the amount of possible consumption in a future period (Hicks, 1946). Business income is defined as the maximum amount that a firm could pay out in current dividends without reducing net worth. This income concept encompasses not only current earnings but also changes in asset levels: Capital gains represent income increase and capital losses income reduction. The depreciation account reflects the fact that unless the capital stock is maintained and replaced, future consumption possibilities will inevitably decline.

Environmental problems may grow progressively worse not only from depletion but also from degradation. If the world economy continues to expand at historical rates, doubling in size every 20–25 years, the biosphere will suffer increasing ecological damage, *unless* the use of resources and the discharge of emissions per unit of output fall as fast as the economy grows. This inescapable fact directs attention to a neglected dimension of economic productivity: productivity in the use of the environment. Conventional measures of productivity change are misleading indicators because they do not account for environmental inputs and outputs, which may be as large as other factors of production.

In this paper we explore two methods for incorporating natural resources into conventional economic performance indicators. The first, natural resource accounting (NRA), is, simply put, a methodology that extends accepted notions of income and depreciation to the stock of natural resources, treating such resources as depreciable assets. The second, multifactor productivity (MP), can be extended to bring in measures of environmental inputs and outputs. The first example provided in this paper looks at adjustments to the national accounts using NRA as a way to measure the economic value of resource depletion in Indonesia. The second example, assessing or analyzing agricultural systems, employs NRA to value soil and water depletion, and also measures off-site damages to account for environmental performance in a larger economic sense. The third example incorporates health costs into MP measures of the electric power industry to arrive at an estimate of productivity change under regulations limiting harmful emissions.

NATIONAL INCOME AND NATURAL RESOURCE ACCOUNTING

The aim of national income accounting is to provide an information framework suitable for analyzing the performance of the economic system. The current system of national accounts reflects the economic concerns that were dominant when the system was developed, particularly the theories of John Maynard Keynes and his contemporaries. The great aggregates of Keynesian analysis—consumption, savings, investment, and government expenditures—are carefully defined and measured. But Keynes and his contemporaries were preoccupied with the Great Depression and the business cycle—specifically, with explaining how an economy could remain for long periods of time at less than full employment. During the Great Depression, commodity prices were at an all-time low. Thus, as Keynesian analysis largely ignored the productive role of natural resources, so does the current system of national accounts.

An earlier generation of classical economists had regarded income as the return on three kinds of assets: natural resources, human resources, and invested capital (land, labor, and capital, in their vocabulary). But natural resource scarcity played little part in nineteenth-century European economics—resources were available and prices were falling. Neoclassical economists from whose work traditional Keynesian and most contemporary economic theories are derived virtually ignored natural resources, focusing on human resources and invested capital. After World War II, when these theories were applied to problems of economic development in the Third World, human resources were also left out on the grounds that labor is always "surplus," and development was seen almost entirely as a matter of savings and investment in physical capital. Ironically, low-income countries, which are typically most dependent on natural resources for employment, revenues, and foreign exchange earnings, are instructed to use a system for national accounting and macroeconomic analysis that almost completely ignores their principal assets.

The result today is a dangerous asymmetry in the way we measure, and hence the way we think about, the value of natural resources. Man-made assets—buildings and equipment, for example—are valued as productive capital. As they wear out, a depreciation charge is taken against the value of production that these assets generate. This practice recognizes that a consumption level maintained by drawing down the stock of capital exceeds the sustainable level of income. But customary accounting methods do not value natural resource assets in this manner: their loss entails no debit charge against current income that would account for the decrease in potential future production. A country could exhaust its mineral resources, cut down its forests, erode its soils, pollute its aquifers, and hunt its wildlife and fisheries to extinction, but measured income would not be affected as these assets disappeared.

The United Nations System of National Accounts (SNA) is the standard framework for measuring a country's macroeconomic performance. The SNA includes stock accounts that identify assets and liabilities at particular points in time and flow accounts that keep track of transactions (e.g., expenditures on goods and services, payments to wage and profit earners, and imports and exports of goods and services) during intervals of time. These national accounts have become the basis for virtually all macroeconomic analysis, planning, and evaluation. Supposedly, they represent an integrated, comprehensive, and consistent framework. Unfortunately, they do not.

Although capital formation is assigned a central role in economic theories, natural resources are not treated like other tangible assets in the system of national accounts. Activities that deplete or degrade natural resources are not recorded as consuming capital, nor are activities that increase the stock of natural resources defined as capital formation. According to the United Nations Statistical Office, ". . . nonreproducible physical assets such as soil or the natural growth of trees . . . are not included in the gross formation of capital, due to the fact that these assets are not exchanged in the marketplace" (United Nations, 1975).

On the other hand, the SNA does classify as gross capital formation expenses incurred in "improving" land for pastures, developing or extending timber-producing areas, or creating infrastructure for the fishing industry. Such actions contribute to recorded income and investment, although they can destroy valuable natural resource assets through deforestation, soil erosion, and overfishing. No record is kept or appears in the national income and investment accounts of this loss of capital as natural resources are used beyond their capacity to recover. The accounts thereby create the illusion of rising national income when in fact national wealth is being depleted.

A Case Study of Indonesia

Indonesia provides an illustration of the potential for NRA, following the depletion method developed in a World Resources Institute (WRI) report (Repetto et al., 1989). Over the past 20 years, Indonesia has drawn heavily on its considerable natural resource endowment to finance development expenditures. Revenues from production of oil, gas, hard minerals, and timber and other forest products have offset a large share of government development and routine expenditures. Resource extraction contributes more than 43 percent of gross domestic product (GDP), 83 percent of exports, and 55 percent of total employment. Indonesia's economic performance from 1965 to 1986 is generally judged to have been successful: its per capita GDP growth averaged 4.6 percent per year, a rate exceeded by only a handful of low- and middle-income countries and far above the average for those groups. Gross domestic investment (GDI) rose from 8 percent of GDP in 1965, at the end of the Sukarno era, to 26 percent of GDP (also well above average) in 1986, despite low oil prices and a difficult debt situation (World Bank, 1988).

Estimates derived from the Indonesian case study illustrate how much this evaluation is affected by keeping score more correctly. Table 1 compares the GDP at constant prices with the net domestic product (NDP), derived by subtracting estimates of net natural resource depreciation for only three sectors: petroleum, timber, and soils. It is clear that conventionally measured GDP substantially overstates net income and its growth rate, because it does not account for consumption of natural resource capital. In fact, although the GDP increased at an average annual rate of 7.1 percent from 1971 to 1984, the period covered by this case study, the adjusted estimate of NDP rose by only 4 percent per year. If 1971, a year of significant additions to petroleum reserves, is excluded, the respective growth rates from 1972 to 1984 are 6.9 percent and 5.4 percent per year for gross and net domestic product, respectively.

The overstatement of income and its growth rate may actually exceed these estimates considerably because the estimates cover only three natural resources—petroleum, timber, and soils—on only one island, Java. Other important exhaustible resources that have been exploited over the period, such as natural gas, coal, copper, tin, and nickel, have not yet been included in the accounts, and neither has the depreciation of such renewable resources as nontimber forest products and fisheries. When complete depreciation accounts are available, they will probably, on balance, show a greater divergence between gross output and net income.

Other important macroeconomic estimates are even more distorted. Table 2 compares estimates of gross and net domestic investment (NDI), the latter reflecting depreciation of natural resource capital. NDI is central to economic plan-

TABLE 1   Comparison of GDP and NDP in 1973 Rupiah (billions)

| | | Net Capital Consumption in Natural Resource Sectors | | | Total Net Capital | |
Year	GDP[a]	Petroleum	Forestry	Soil	Consumption	NDP
1971	5,545	1,527	−312	−89	1,126	6,671
1972	6,067	337	−354	−83	−100	5,967
1973	6,753	407	−591	−95	−279	6,474
1974	7,296	3,228	−533	−90	2,605	9,901
1975	7,631	−787	−249	−85	−1,121	6,510
1976	8,156	−187	−423	−74	−684	7,472
1977	8,882	−1,225	−405	−81	−1,711	7,171
1978	9,567	−1,117	−401	−89	−1,607	7,960
1979	10,165	−1,200	−946	−73	−2,219	7,946
1980	11,169	−1,633	−965	−65	−2,663	8,506
1981	12,055	−1,552	−595	−68	−2,215	9,840
1982	12,325	−1,158	−551	−55	−1,764	10,561
1983	12,842	−1,825	−974	−71	−2,870	9,972
1984	13,520	−1,765	−493	−76	−2,334	11,186
Percent average annual growth	7.1					4.0

[a]From the Indonesian Central Bureau of Statistics.

SOURCE: Repetto et al. (1989).

ning in resource-based economies. Countries such as Indonesia that are heavily dependent on exhaustible natural resources must diversify their asset base to preserve a sustainable long-term growth path. Extraction and sale of natural resources must finance investments in other productive capital. It is therefore relevant to compare the figures for GDI with those representative of natural resource depletion. If gross investment is less than resource depletion, the country is drawing down, rather than building up, its asset base and using its natural resource endowment to finance current consumption. If net investment is positive, but not enough to equip new workers with at least the capital per worker of the existing labor force, then increases in output per worker and income per capita are unlikely. In fact, the results from the Indonesian case study show that the adjustment for natural resource asset changes is large in many years relative to GDI. In 1971 and 1974, the adjustment is positive, due to additions to petroleum reserves.[1] In most years during the period, however, the depletion adjustment offsets a good part of gross capital formation. In some years, net investment was negative, implying that natural resources were being depleted to finance current consumption expenditures.

Such an evaluation should flash an unmistakable warning signal to economic policy makers that they are on an unsustainable course. An economic accounting system that does not generate and highlight such evaluations is deficient as a tool for analysis and policy in resource-based economies.

The same holds true for the evaluation of performance in particular economic sectors, such as agriculture. Almost three-quarters of the Indonesian population lives on the fertile but overcrowded inner islands of Java, Bali, and Madura, where lowland irrigated rice paddies are intensively farmed. In the highlands, population pressures have brought steep hillsides into use for cultivation of maize, cassava, and other annual crops. As hillsides have been cleared of trees, erosion has increased, to the point where it now averages over 60 tons per hectare per year, by WRI estimates.

Erosion's economic consequences include loss of nutrients and soil fertility as well as increased downstream sedimentation in reservoirs, harbors, and irrigation systems. Increased silt concentrations affect fisheries and downstream water users. Although crop yields have improved in the hills because farmers have used better seed and more fertilizers, estimates indicate that the annual depreciation of soil fertility (calculated as the value of lost farm income) is about 4 per-

TABLE 2   Comparison of Gross Domestic Investment and Net Domestic Investment in 1973 Rupiah (billions)

Year	GDI[a]	Resource Depletion[b]	NDI
1971	876	1,126	2,002
1972	1,139	−100	1,039
1973	1,208	−279	929
1974	1,224	2,605	3,829
1975	1,552	−1,121	431
1976	1,690	−684	1,006
1977	1,785	−1,711	74
1978	1,965	−1,607	358
1979	2,128	−2,219	−91
1980	2,331	−2,663	−332
1981	2,704	−2,215	489
1982	2,783	−1,764	1,019
1983	3,776	−2,870	906
1984	3,551	−2,334	1,217

[a]From the Indonesian Central Bureau of Statistics.
[b]Includes depletion of forests, petroleum, and the cost of erosion on the island of Java.

SOURCE: Repetto et al. (1989).

cent of the value of crop production—the same percentage as annual production increases. In other words, these estimates suggest that *current* increases in farm output in Indonesia's uplands are being achieved almost wholly at the expense of decreases in *future* output. Because the upland population is unlikely to be smaller in the future than it is now, soil erosion represents a transfer of wealth from the future to the present. By ignoring the future costs of soil erosion, the sectoral income accounts significantly overstate the growth of agricultural income in Java's highlands.

NATURAL RESOURCE ACCOUNTING AND AGRICULTURE

In recent years, a number of researchers have struggled to define sustainable agriculture. Most of these definitions encompass elements of agricultural productivity maintenance, farm profitability, and reduction of environmental impacts, but they have been qualitative, not quantitative. Also, most definitions of agricultural sustainability have failed to incorporate productivity of the natural resource base when calculating agricultural productivity. The notion of agricultural sustainability has therefore been of considerable conceptual utility but only limited operational usefulness to policy makers and researchers attempting to determine how various policies and technologies affect agricultural resources.

A Natural Resource Accounting Framework for Agriculture

An NRA framework differs from conventional financial and economic accounting in some significant ways (Faeth, 1993). In conventional accounting, the *financial* value (net farm income) of a production program to farmers takes into account current and future transfer receipts but ignores environmental costs. Using the NRA framework, net farm income is defined to include the value of changes in soil productivity, the farmer's principal natural asset. This definition is consistent with business and economic accounting standards, which incorporate asset formation and depreciation in measures of income. By contrast, the same farming technique's *economic* value to society (net economic value) includes environmental costs that farmers' activities impose on others, such as damages related to surface water, but ignores transfer payments.

Tables 3 and 4 present examples of this NRA methodology. The tables compare net farm income and net economic value per acre for a predominantly corn-soybean rotation in Pennsylvania, with and without allowances for natural resource depreciation. Column 2 of Table 3 shows a conventional financial analysis of net farm income per acre per year. The gross operating margin ($75) (crop sales less variable production costs) is shown in the first row. Because conventional analyses make no allowance for natural resource depletion, the gross margin and net farm operating income are the same. Government subsidies ($16) are added to obtain net income ($91).

TABLE 3 Net Farm Income: Conventional versus Natural Resource Accounting (dollars per acre per year)

Item Accounting	Conventional Accounting	Natural Resource
Gross operating margin	75	75
Less soil depreciation	—	24
Net farm operating income	75	51
Plus government commodity subsidy	16	16
Net farm income	91	67

When a soil depreciation allowance is included, the gross operating margin is adjusted ($24) to obtain net farm income ($51). The depreciation allowance is an estimate of the present value of future income losses due to the impact of crop production on soil quality. The same government payment is added to determine net farm income ($67).

Net economic value (Table 4, column 3) subtracts $49 as an adjustment for off-site costs of soil erosion (such as sedimentation, impacts on recreation and fisheries, and effects on downstream water users).[2] Net economic value also includes the on-site soil depreciation allowance, but excludes income support payments. Farmers do not bear the off-site costs directly, but these are real economic costs attributable to agricultural production and should be considered in calculating net economic value to society. Subsidy payments, by contrast, are a

TABLE 4   Net Economic Value: Conventional versus Natural Resource Accounting (dollars per acre per year)

Item Accounting	Conventional Accounting	Natural Resource
Gross operating margin	75	75
Less soil depreciation	—	24
Net farm operating income	75	51
Less off-site costs of soil erosion	—	49
Net farm income	75	2

transfer from taxpayers to farmers, not income generated by agricultural production, and are therefore excluded from net economic value calculations.

Case Studies

WRI has published a series of six case studies that explicitly examined sustainable agricultural practices, including on- and off-farm economic measures of agricultural sustainability (Faeth et al., 1991; Faeth, 1993). These include two case studies of alternative corn–soybean production systems in Pennsylvania and Nebraska, rice–wheat–maize production systems in India, lowland irrigated rice in the Philippines, and a comparison of upland and lowland wheat production in Chile. Each study is based on actual field trials.

These studies applied an NRA framework to quantify the financial, economic, fiscal, and environmental costs and benefits of various agricultural practices. Within this framework, we accounted for the value of long-term soil productivity changes and off-site surface water damages for alternative farming practices. We also analyzed the financial value to farmers and the economic value to society of each farming practice under five policy scenarios.

In the Pennsylvania case study, organic farming practices proved superior to conventional practices agronomically, environmentally, and economically. Resource-conserving production practices cut production costs by 25 percent, eliminated chemical fertilizer and pesticide use, reduced soil erosion by more than 50 percent, and increased yields after completion of a transition from heavy chemical use. In addition, increasing water retention reduced off-site damages by $30 per acre per year, and reducing erosion forestalled a 30-year yield decline with a present value of more than $124 per acre.

In Nebraska, low-chemical-input alternatives to the predominant corn–soybean rotation were found to be economically competitive and environmentally superior. Three different regimens for the corn–soybean rotation (herbicide and fertilizer use, fertilizer use only, and an organic treatment) yielded farm incomes and net economic values that differed by no more than $2 per acre per year.

In northwest India, heavy electricity subsidies for tubewell irrigation are resulting in the depletion of groundwater at the rate of 0.8 meter per year. The value of this loss in terms of future pumping costs represents almost 15 percent of gross operating margin, and when it is included in financial calculations, water-conserving farming practices are seen to be much more profitable. In the Philippines, when the health-care costs for farmers who apply unregulated, subsidized pesticides are accounted for, scheduled spraying of pesticides is much less profitable than integrated pest management or biocontrol methods. And in Chile, where poor farmers use soil-degrading production practices on steep hillsides, soil-conserving practices are more profitable than traditional methods.

Several important conclusions emerged from this research:

• Economic analysis that excludes the value of productivity changes of natural resources or externalities will overstate the value of resource-degrading practices and understate the value of resource-conserving practices.

• Resource-conserving production practices can be economically and financially superior to, or competitive with, conventional practices.

• Failure to account for the degradation and depletion of natural resources can mask their true economic value, thus justifying policies that diminish sustainability and result in significant economic and fiscal losses.

A Sectoral Study of Agriculture

WRI completed a major national economic analysis of agricultural sustainability in the United States in 1995. This study applied an NRA framework to analyze the economic and environmental impacts of alternative policies and production systems (Faeth, 1995). After compiling agronomic data from experiments, field trials, and producer records for alternative production systems in 10 regions of the United States, WRI evaluated these alternative systems, as well as the predominant systems for a given region, using a biophysical soil and crop model to determine soil erosion rates, long-term crop yields, nutrient loss, potential groundwater contamination from nitrates, and soil carbon sequestration. This information, together with financial and energy analysis and economic valuation of environmental impacts, makes the database supporting this project the single most complete collection of information yet available on "sustainable" production systems.

The economic model that resulted from the WRI study is the most comprehensive and empirically based policy tool yet developed for analysis of agricultural sustainability in the United States. To date, no national economic model has used such extensive information on alternative production systems, the environmental impact of predominant and alternative farming systems, or the economic value of natural resource impacts.

The research plan involved four steps:

• collecting and organizing existing agronomic data on predominant and alternative[3] production systems,

• calculating crop budgets and simulating the environmental characteristics for each predominant and alternative production system,

• reprogramming an existing economic policy model to incorporate alternative production systems and physical and economic accounts for natural resource impacts of both predominant and alternative systems and establishing an economic baseline for the adapted model, and

• using the adapted model to test alternative policy scenarios and undertake sensitivity analysis.

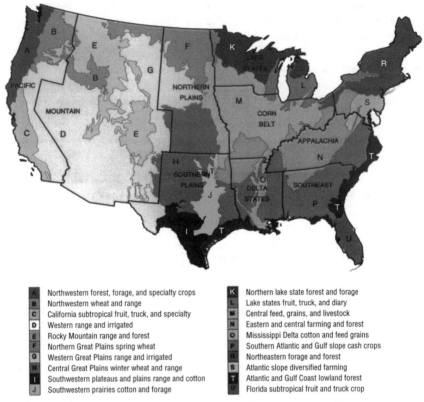

A	Northwestern forest, forage, and specialty crops	**K**	Northern lake state forest and forage
B	Northwestern wheat and range	**L**	Lake states fruit, truck, and diary
C	California subtropical fruit, truck, and specialty	**M**	Central feed, grains, and livestock
D	Western range and irrigated	**N**	Eastern and central farming and forest
E	Rocky Mountain range and forest	**O**	Mississippi Delta cotton and feed grains
F	Northern Great Plains spring wheat	**P**	Southern Atlantic and Gulf slope cash crops
G	Western Great Plains range and irrigated	**R**	Northeastern forage and forest
H	Central Great Plains winter wheat and range	**S**	Atlantic slope diversified farming
I	Southwestern plateaus and plains range and cotton	**T**	Atlantic and Gulf Coast lowland forest
J	Southwestern prairies cotton and forage	**U**	Florida subtropical fruit and truck crop

FIGURE 1   Land resource regions (LRRs) as of January 1984. SOURCE: Faeth (1995).

The database developed for this project encompasses predominant and alternative production systems for major crops in each of the 10 U.S. Department of Agriculture (USDA) production regions (Figure 1). Crops covered in the database include corn, sorghum, barley, oats, wheat, rice, soybeans, cotton, and hay. All alternative systems that our agronomic team could identify for which experimental or field data exist were included in the database. Predominant systems were identified using the Cropping Practices Survey (Daberkow and Gill, 1989) developed by the National Agricultural Statistical Service and Economic Research Service (ERS) and the Farm Costs and Returns Survey developed by the National Agricultural Statistical Service.

The data collected in the course of this study included basic agronomic data such as crop yield, input use, crop sequence, and field operations. These data provided a solid foundation for deriving estimates of various characteristics of each farming system, including cost of production, soil erosion rates, leachate contamination, and soil carbon sequestration. All data or estimates used for this analysis were based on experimental or field data and the USDA's Erosion Productivity Impact Calculator (EPIC) (Williams and Renard, 1985).

As with WRI's case studies, the USDA's EPIC model was used to estimate soil erosion rates, short- and long-term crop yields, nutrient runoff, potential groundwater contamination, and soil carbon sequestration for each production system in each region. These representative estimates were based on the principal land resource regions (LRRs) for each of the 10 U.S. production regions and the predominant soils in those LRRs (Figure 1). The analysis was disaggregated into 48 LRRs for agronomic and environmental evaluation.[4]

The U.S. Math Programming (USMP) model (House, 1987), developed by the ERS over the past decade for national economic policy analysis, was adapted for use in this study. In collaboration with ERS, WRI produced an NRA version of the USMP model by extending it to include alternative commodity production systems, soil depreciation allowances, soil carbon sequestration, energy budgets, and regional natural resource damages (Figure 2). Prior to this effort, the model covered predominant production practices only and did not include any environmental impacts.

Using the completed NRA version of the USMP model, WRI tested a variety of agricultural policy scenarios for the 1995–1996 farm bill discussions. The analysis estimated several variables for each policy scenario for each region, including commodity production, commodity prices, farm income, net economic value of agricultural production, fiscal cost of income support, value and level of agricultural trade, land use, gross soil erosion, value of soil depreciation, value of soil carbon sequestered, and value of off-farm surface water impacts.

The analysis was done from the standpoint of maximizing farmers' incomes over the long term, with postsolution calculations of public welfare (Chandler et al., 1981), because farm production decisions are made by farmers, not policy makers. To estimate the value of production to the farmer, we calculated net farm income, incorporating gross operating margin, a soil depreciation allowance for changes in soil productivity, and commodity program payments. The value of off-site resource damages was excluded because farmers do not pay these.

Public welfare was estimated by calculating the net economic value of production, using gross operating margin, a soil depreciation allowance, the value of off-site surface water damages (because society absorbs the costs of these damages), and the value to society of soil carbon sequestration (because mitigating global warming benefits society as a whole). Income support was excluded because these transfer payments do not alter the net economic value.

Through the analysis described above, the revised USMP model can identify the optimal technologies for each policy scenario and estimate their potential extent of use as determined by relative profitability. In this way, estimates can be generated of the physical extent and economic value of natural resource impacts for a given policy and technology mix.

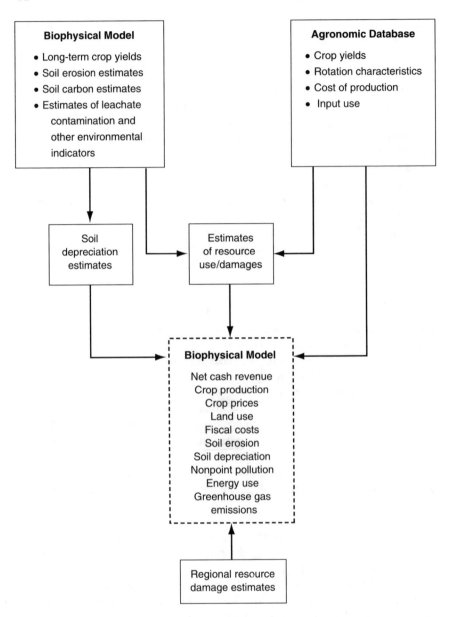

FIGURE 2   Modifications to incorporate resource costs into USDA's U.S. Math Programming model for economic policy analysis. SOURCE: Faeth (1995).

The adapted model was first used to develop a standard baseline scenario reflecting the Food, Agriculture, Conservation and Trade Act of 1990 and production practices as of 1992, the last year for relevant national surveys. This baseline was then extended to take into account alternative production practices and long-term changes in soil productivity. Finally, the extended model was used to test a variety of policies, including different levels of flexibility and green-payment options for improving the environmental performance of agricultural policy.

The results of this study demonstrate that a reduction in agriculture's impact on the environment is both possible and economically advantageous. Only alternative production practices that improve farmers' bottom line are assumed to be adopted, and the alternatives represented are being used by small numbers of farmers or being developed and field tested by agricultural scientists. Many of these practices reduce production costs by improving input-use efficiency. Such alternative production practices could greatly help farmers conserve resources, improve productivity and profits, and reduce fiscal costs. Policy changes that remove the biases against such production practices would allow further improvement and save taxpayers money.

The extended baseline scenario implies that if farmers fully accounted for the cost of long-term soil productivity changes and if alternative production practices were fully available, soil erosion and its off-site costs would go down significantly. In the extended baseline, soil erosion is reduced nationally by 7 percent and damages by 10 percent. Soil depreciation cost estimates are relatively small, compared with other calculated production costs, and they are actually negative for production practices leading to yields estimated to increase over time. The larger effect in the baseline extension comes from including the alternative practices in the set of practices that the model can choose from. Many alternative practices turn out to be very competitive financially and come into the extended baseline model solution based solely on relative profits even though they also provide environmental benefits.

Policy analysis showed that green-payment options to subsidize conservation also help to improve environmental performance, but not all green-payment programs would work equally well, environmentally or fiscally. Targeted subsidies adjusted to account for the value of regional damages achieve lower program costs and have the greatest benefits.

Environmental performance increases with increasing commodity program flexibility. For the scenario we tested with the highest degree of flexibility, the indicators of environmental performance were nearly the same as for the best green-payment case. Two changes account for the environmental results under increased flexibility: both the acreage in production and the use of monocultural practices decline as flexibility increases.

PRODUCTIVITY MEASUREMENT IN THE
ELECTRIC POWER INDUSTRY

Productivity growth has long been an important concern of economists, industrialists, and government officials because it is recognized as the key to business profitability and economic welfare. The apparent lag in productivity growth in the U.S. economy, relative to other industrial countries and to our own past record, has generated many diagnoses and diverse policy prescriptions.

These concerns emerge from conventional measurements of productivity change that encompass only marketed outputs and inputs. Labor productivity, for example, measures output per worker. A broader measure, multifactor productivity (MP) (sometimes also called total factor productivity), measures output per unit of an index of labor, capital, and intermediate materials inputs. In this analytical framework, productivity change is defined as the difference between the growth rate of output and that of the index of inputs.

Almost no attempt has been made to measure environmentally related outputs (such as emissions) or inputs (such as natural resource services) that are not marketed or to assess their significance for economic productivity. What follows is an exploratory step in that direction, using the private electric power industry in the United States as an example. (For an update of this study and two additional cases including the pulp and paper and agriculture sectors, see Repetto et al., 1996.)

A typical 500-MW coal-fired power plant produces more than just 3.5 billion kWh of electricity per year. The 1.5 million tons of coal and 0.15 million ton of limestone it uses as inputs reappear in some form as outputs. Emissions to the atmosphere include 1 million tons of carbon in the form of carbon dioxide, 5,000 tons of sulfur as sulfur dioxide, 10,000 tons of nitrogen oxides formed largely from air drawn into the combustion process, and a variety of other compounds. Solid outputs include 140,000 tons of ash and 193,000 tons of scrubber sludge, which contain 5,000 and 40,000 tons of sulfur, respectively.

A more general measure of economic productivity, recognizing the conservation of matter and energy, would assess the extent to which the industrial transformation of materials has yielded outputs with greater economic benefits—or lower economic costs—than the costs of the inputs. Many of the power plant's unmarketed outputs have economic significance. Airborne sulfur emissions, for example, affect human health, plant growth, and the durability of materials. As recent experiments with marketable emissions rights indicate, the extent to which such outputs are marketed is largely an institutional arrangement. Productivity measures restricted to a subset of economically significant inputs and outputs can misrepresent technological progress in the industry.

As a step toward a broader measure, we developed an index of atmospheric emissions by weighting each of three pollutants (sulfur and nitrogen oxides and particulate matter) according to their estimated economic significance, defined as the marginal damages caused by an additional ton emitted in 1987. Kilowatt hours generated per unit of emissions, a single-factor measure of environmental productivity, improved much more rapidly than capital, labor, or energy productivity between 1970 and 1987. The index of emissions fell by 30 percent over this period, while electricity output increased. Had emissions per kilowatt hour remained at the 1970 level, the electric power industry would have emitted 10 million more tons of sulfur dioxide in 1987 than it did.

If the electric utility produces both kilowatt hours and emissions, a MP measure can be constructed incorporating the changing output mix. The share of emissions in total output should be measured in economic terms using their "shadow prices" (i.e., estimates of the marginal damages to the economy occasioned by an additional ton of each pollutant). Such estimates are hard to obtain, because damages vary across the country and are not reflected in market transactions. As the best available approximations, we used estimates prepared by U.S. Environmental Protection Agency and reviewed by the Office of Management and Budget. Their best estimate of marginal damage costs for sulfur dioxide emissions in 1987 is $637 per ton, with a range of $290 to $1,612. Damages to health, materials, agriculture, and visibility are included in the total, but contributions to acidic deposition or climate change are not. Analogous numbers for particulate matter and nitrogen oxides are $2,550 and $230 per ton, respectively.[5]

Using these shadow prices to estimate the value of emissions in each year,[6] the cost of emissions as unpriced outputs of the electric power industry in the 1980s was about as large as the cost of labor to the industry. Hence, emissions have as great a weight as labor in a generalized productivity index.

In summary, productivity growth in the private electric power industry during these years appears two to three times as high if its progress in reducing economically damaging emissions is taken into account. In addition, productivity in the industry increased more rapidly in the 1970s, when emissions were being reduced more rapidly, than it did in the 1980s, when the rate of decline was more modest. This is contrary to the conclusions of conventional MP measurements, which do not incorporate environmental imports (Table 5).

Although this is a preliminary exploration, it suggests that technologies that reduce environmental damages contribute significantly to economic productivity. They do not merely raise production costs. The example also suggests that it is important to measure environmental dimensions of productivity to avoid one-sided assessments.

CONCLUSION

Economic indicators and analyses that count the cost of environmental protection but ignore the cost of environmental degradation and the loss of natural assets mislead both public and private decision makers. The problems with productivity measurement, for example, have led to serious misunderstandings about

TABLE 5   Estimates of Multifactor Productivity in
the Electric Power Industry (percent per year)

		Incorporating Emission	
Period	Conventional Measure	Index A[a]	Index B[b]
1971–1985	−0.38	0.62	0.33
1971–1979	−1.10	0.26	−0.18
1980–1985	0.69	1.17	1.10

[a]Index A assumes that marginal damages were constant in real
terms. Refer to footnote 6 for more information.
[b]Index B assumes that marginal damages increased in real terms in
proportion to gross national product. Refer to footnote 6 for more
information.

SOURCES: Jorgenson and Fraumeni (1990) and U.S. Environmental Protection Agency (1994).

the effects of environmental policies on the economy and distortions in the policy-
making process. As productivity declined during the 1970s, economic studies
claimed that environmental regulations were responsible for up to half of the
productivity decline observed in pollution-intensive industrial sectors.

This and similar findings continue to resonate in current environmental policy
debates. Behind efforts to weaken environmental laws or their enforcement lies
the belief that such regulations impose costly burdens on the economy, stifling
innovation and lowering productivity. However, the conclusion that environmental regulations have reduced the rate of productivity growth is an artifact of a
basic flaw in the way productivity is measured, as the case presented here for the
electricity sector demonstrates.

Similarly, the failure to include natural resource stocks in national accounts
leads to the mistaken notion that their depletion contributes to income growth.
Again, this follows from a key omission in an important economic indicator.

The series of studies reviewed here show that when economic indicators are
restructured to include environmental gains and losses, the results will lead to
conclusions that support the economic efficacy of environmental policy: pollution control can increase productivity by reducing environmental damages; wise
use of natural resources makes economic sense; and the elimination of commodity subsidies is fiscally as well as environmentally sound.

NOTES

1.   It may seem anomalous that in 1971 and 1974 depreciation was a negative number, that is, net
     capital consumption was added to GDP and investment. The reason for this is that the value of

additions to petroleum reserves in these years was considerably larger than all categories of depletion combined, leading to negative depreciation.

One way of resolving this apparent anomaly would be to account separately for additions and subtractions from natural resource assets. Real capital gains (as distinct from those resulting from price changes) can be accounted for as gross income and gross capital formation. This is consistent with our earlier definition of income because additions to resources during the current year augment the amount that could be consumed currently without reducing potential consumption in future years. This is obvious in the case of forest growth but less obvious for mineral discoveries, because current discoveries may leave less to be discovered later on. However, insofar as additions to mineral reserves reflect advances in the technology of exploration or extraction, the total potential resource base will have expanded.

2. The value used for off-site damages from soil erosion in the Northeast is $8.16 per ton of eroded soil. Because these numbers were calculated (Ribaudo, 1989) based on gross erosion and gross damages, sediment delivery need not be estimated. Values are available for each production region.

3. Our working definition of "alternative" includes those production practices that enhance environmental quality and make efficient use of nonrenewable resources. This follows the legal definition of "sustainable." Thus, some practices that may be considered "conventional," such as reduced tillage, may be included as alternatives if they are not the predominant practice in a given region.

4. There are 20 LRRs identified by the Natural Resources Conservation Service in the 48 contiguous states. Few LRRs are contained within a single production region. Where an LRR is cut by a production region, we have split the LRR. This analysis does not include Alaska or Hawaii.

5. The U.S. Envronmental Protection Agency's damage estimate for nitrogen oxides included a "credit" for a reduction in smog formation with increasing NO_x emissions. This credit was omitted in the analysis because the NO_x's smog-inhibiting effect is temporary and spatially limited. The marginal damage cost, including the credit, would be $69 per ton.

6. Because year-by-year estimates of marginal damages were not available, two alternative assumptions were used to extrapolate the marginal damages backward in time to 1970. The assumptions reflect two offsetting trends: in earlier years, emissions were greater, so marginal damages should have been higher, but the size of the economy was smaller, so that damages should have been smaller, also. The first assumption, therefore, was that marginal damages were constant over the period in real terms (Index A of Table 5). The alternative was that marginal damages increased in real terms in proportion to real GNP (Index B of Table 5). Alternative estimates of MP were made reflecting these assumptions.

REFERENCES

Chandler, W., J. Fortuny-Amat, and B. McCarl. 1981. The potential role of multilevel programming in agricultural economics. American Journal of Agricultural Economics 63(1):521–531.

Daberkow, S., and M. Gill. 1989. Common crop rotations among major field crops. Pp. 34–39 in Agricultural Resources: Inputs Situation and Outlook Report. Report No. AR-15. Washington, D.C.: Economic Research Service, U.S. Department of Agriculture.

Faeth, P., ed. 1993. Agricultural Policy and Sustainability: Case Studies from India, Chile, the Philippines and the United States. Washington, D.C.: World Resources Institute.

Faeth, P. 1995. Growing Green: Enhancing the Economic and Environmental Performance of U.S. Agriculture. Washington, D.C.: World Resources Institute.

Faeth, P., R. Repetto, K. Kroll, Q. Dai, and G. Helmers. 1991. Paying the Farm Bill: U.S. Agricultural Policy and the Transition to Sustainable Agriculture. Washington, D.C.: World Resources Institute.

Hicks, J. R. 1946. Value and Capital: An Inquiry into Some Fundamental Principles of Economic Theory. Oxford: Oxford University Press.

House, R. 1987. USMP Regional Agricultural Model (GAMS version guide), Unpublished mimeo, Economic Research Service, Washington, D.C.

Jorgenson, D., and B. Fraumeni. 1990. Productivity and U.S. Economic Growth: 1979–1985, unpublished discussion paper for the Harvard Institute of Economics, Harvard University, Cambridge, Mass.

Repetto, R., W. Magrath, M. Wells, C. Beer, and F. Rossini. 1989. Wasting Assets: Natural Resources in the National Income Accounts. Washington, D.C.: World Resources Institute.

Repetto, R., D. Rothman, P. Faeth, and D. Austin. 1996. Has Environmental Protection Really Reduced Productivity Growth? We Need Unbiased Measures. Washington, D.C.: World Resources Institute.

Ribaudo, M. 1989. Water Quality Benefits from the Conservation Reserve Program. Agricultural Economic Report No. 606. Washington, D.C.: Resources and Technology Division, Economic Research Service, U.S. Department of Agriculture.

United Nations. 1975. A System of National Accounts. New York: United Nations.

U.S. Environmental Protection Agency. 1994. National Air Pollution Emission Trends, 1900–1993. Washington, D.C.: Office of Air Quality Planning and Standards, U.S. Environmental Protection Agency.

Williams, J. R., and H. G. Renard. 1985. Assessments of soil erosion and crop productivity with process model (EPIC). Pp. 68–102 in Soil Erosion and Crop Productivity, R. F. Follett and B. A. Stewart, eds. Madison, Wis.: American Society of Agronomy.

World Bank. 1988. World Development Report. Oxford: Oxford University Press.

Measures of Environmental Performance and
Ecosystem Condition. 1999. Pp. 89–95.
Washington, DC: National Academy Press.

Environmental Performance Standards for Farming and Ranching

CRAIG COX AND SUSAN E. OFFUTT

Farming and ranching can degrade soil, pollute groundwater and surface water, and disrupt terrestrial and aquatic ecosystems. The effort to develop farming and ranching systems that sustain the natural resource base and reduce the adverse effects of agricultural production on the environment has highlighted the need for better measures of resource condition and, ultimately, for indicators of environmental performance.

Two National Research Council (NRC) reports address the scientific basis for constructing condition and performance measures for farming and ranching. *Soil and Water Quality: An Agenda for Agriculture* (1993a) considers the biological, chemical, and physical processes that determine the effect of farming on soil and water quality. The report emphasizes the fundamental importance of soil and the linkage among soil, water quality, and water pollution. It also proposes a set of measurable criteria as a starting point for a more comprehensive set of environmental performance standards for farming systems. *Rangeland Health: New Methods to Classify, Inventory, and Monitor Rangelands* (1994) recognizes that the productivity of extensively managed systems such as rangelands depends on the maintenance of the integrity of soil and ecological processes. The report contains criteria and indicators for assessing whether the productive capacity of a rangeland is being sustained. These NRC efforts represent steps in the evolution of performance standards for farm and ranch management.

Each study took a functional approach to the development of standards. In the soil and water study, the NRC committee identified three primary functions carried out by these two resources: (1) promotion of plant growth, (2) regulation and partition of water through watersheds, (3) and buffering the effect of agricultural chemicals or other inputs to production on the environment. In the rangeland study, the committee considered primarily one resource function: the production of different kinds, amounts, and arrangements of vegetation. The characterization of resource functions is mostly a scientific endeavor, but the development of standards requires explicit value judgments. So, although soils perform many functions, the three selected as a basis for measuring performance were those considered most important for sustaining agricultural productivity and protecting water quality. The selection of performance standards depends, fundamentally, on the economic and noneconomic values placed on the use and existence of the natural resource in question.

SOIL QUALITY

As embodied in such laws as the Clean Air Act and the Clean Water Act, national policy has recognized the importance of air and water quality to the country's well-being. However, there is no equivalent federal "Soil Quality Act," despite the critical role soil plays in mediating both water and air quality. *Soil and Water Quality* urges that soil quality be a national environmental priority:

> The quality of a soil depends on attributes such as the soil's texture, depth, permeability, biological activity, capacity to store water and nutrients, and the amount of organic matter contained in the soil. Soils are living, dynamic systems that are the interface between agriculture and the environment. High-quality soils promote the growth of crops and make farming systems more productive. High-quality soils also prevent water pollution by resisting erosion, absorbing and partitioning rainfall, and degrading or immobilizing agricultural chemicals, wastes, or other potential pollutants. (National Research Council, 1993a, p. 2)

Traditionally, soil quality has been equated with soil productivity, a measure of promotion of plant growth. Soils perform a much broader range of functions in the environment, however, including regulation of water flow in watersheds and of greenhouse gas emission, attenuation of natural and artificial wastes, and regulation of air and water quality (National Research Council, 1993a). Consequently, measures of soil quality will have to be altered to reflect these aspects.

No comprehensive index of soil quality yet exists that captures fully soil's function in the ecosystem. *Soil and Water Quality* notes that it would be "impossible and unnecessary to monitor changes" in all of the soil attributes that relate to critical ecosystem functions (National Research Council, 1993a, p. 205). Moreover, the set of relevant indicators can be expected to change with geographic variation in soil types. The report does suggest a set of indicators that includes the most relevant physical, chemical, and biological attributes of the soil: nutrient availability, organic carbon, labile carbon, texture, water-holding capacity, soil structure; maximum rooting depth, salinity, and acidity/alkalinity (National Research Council, 1993a, p. 208). The report further recommends the development of "pedotransfer functions" that could be used to link quantitatively the measurement of soil quality to the functions soils perform. Efforts are under way to develop such an index, but these will entail a significant amount of research. National assessments of soil resources are conducted currently, but the kind of data collected and the approaches used to analyze the data (which focus largely on descriptions of soil types and on assessments of soil loss) do not facilitate a comprehensive assessment of soil quality.

WATER QUALITY

The difficulties involved in developing performance standards for water quality are even more daunting than those related to standards for soil quality. Numerical criteria for water quality are set by regulatory agencies primarily to protect human health. Numerical criteria or other indicators related to the function of aquatic ecosystems, however, are not as well developed. Another recent NRC report, *Restoration of Aquatic Ecosystems* (1993b), stresses the need to develop both structural and functional criteria to assess the success of aquatic restoration projects. The absence of criteria for aquatic ecosystems makes it difficult to develop performance standards for farming and ranching. This problem is compounded by the difficulty in linking pollutants leaving a particular farm or ranch to their effect on the environment. For example, what level of total dissolved nitrate entering a tributary to the Susquehanna River from an adjacent dairy farm threatens water quality in the Chesapeake Bay?

The difficulty of quantifying these linkages has led researchers to propose qualitative standards that could be applied in more systematic ways to farming and ranching systems. The NRC soil and water quality report proposed two such criteria: the efficiency with which pesticides, nutrients, and irrigation water are used in farming systems; and the degree to which farming systems resist erosion and runoff. The efficiency criterion addresses the inputs to the farming system, whereas the resistance criterion addresses outputs. Quantitative measures of input efficiency and of erosion resistance are available and have been incorporated in models that predict the delivery of agricultural chemicals and sediment to groundwater or surface water. These measures are useful for indicating progress toward alleviating environmental degradation by farming, but the question of how much improvement has occurred remains unresolved. Agreement on nonpoint-source control provisions in the reauthorization of the Clean Water Act has been hampered by such uncertainty.

Soil and Water Quality recognizes that the inability to relate changes in farming practices to changes in soil and water quality will ultimately hamper attainment of environmental goals. The lack of quantifiable performance standards prevents the evaluation of integrated farm plans, plans that can address the range of soil functions and avoid the inconsistencies that can result from a focus on single best-management practices.

> Current understanding of the effect of farming systems on soil and water quality is generally sufficient to identify the best available production practices or management systems; it is not, however, sufficient for making quantitative estimates of how much soil and water quality will improve as a result of the use of alternative practices or management methods. (National Research Council, 1993a, p. 11)

THE HEALTH OF RANGELAND

The debate over the use of public grazing lands is a manifestation of the larger issue of performance standards for managed ecosystems. Indeed, much of the controversy over Interior Secretary Bruce Babbit's proposal for rangeland reform centers on the development and application of "standards and guidelines" for rangeland managed by federal agencies. *Rangeland Health* observes that overgrazing, drought, erosion, and other human and naturally induced stresses have resulted in degradation in the past, though the "present state of health of U.S. rangelands is a matter of sharp debate" (National Research Council, 1994, p. 1).

Diverse rangeland ecosystems produce both tangible commodities with economic value (forage for livestock, for example) and intangible products, such as natural beauty and wilderness, that may have economic value but that also satisfy other important societal values. Protection of the capacity of rangelands to produce commodities and satisfy societal values is the congressionally established mandate for federal range management. In its report, the NRC committee attempted to identify criteria that could be used to monitor that capacity. As contrasted with cropped farming systems, which to greater or lesser extent depend on external inputs such as fertilizers, rangelands do not generally receive such supplements.

> The capacity of rangelands to produce commodities and satisfy societal values depends on the integrity of internal nutrient cycles, energy flows, plant community dynamics, an intact soil profile, and stores of nutrients and wastes. (National Research Council, 1994, p. 5)

The report defines rangeland health as "the degree to which the integrity of the soil and the ecological processes of rangeland ecosystems are sustained" (National Research Council, 1994, p. 2), and it argues for the establishment of a minimum standard of rangeland management that would protect against human-induced loss of rangeland health. This minimum is to be an ecological standard, independent of the rangeland's use and how it is managed, recognizing that if its health is preserved, the rangeland could accommodate a variety of uses (including livestock production and recreation, for example). As does the NRC report on soil and water quality, the rangeland report emphasizes that measures of condition would not be sufficient to guide decisions about uses and management practices, and it notes the need for other data and for aggregate assessments of rangeland health at the national level.

The committee recommended three criteria for making a determination of the state of rangeland health: the degree of soil stability and watershed function, which is critical to the prevention of soil degradation; nutrient cycling and energy flow; and the ability of the rangeland to adapt to change, which is necessary to maintain or move toward a healthy state and might be indicated by increases in vegetative cover or changes in plant age–class distributions.

Although it is not specific about how to quantify and combine indicators relevant to each criterion, the report does present an evaluation matrix that relates indicators to categories of ecosystem health. For example, the distribution and incorporation of plant litter in the soil could be used to assess the degree of nutrient cycling. Declines in production of plant matter and consequent reduction in the incorporation of plant litter into the soil may occur because of overgrazing by livestock. Such outcomes would indicate a diminution of the total volume of nutrients in the rangeland ecosystem.

STATE-OF-THE-ART MEASURES OF
CONDITIONS AND PERFORMANCE

Even this cursory review of the findings of recent NRC reports on soil quality and rangeland health is sufficient to confirm the relatively undeveloped state of quantitative measures of environmental condition and performance for agriculture. That is not to say the lack of good measures should constrain immediate efforts to improve the environmental sensitivity of management practices. To the contrary, both reports address the current possibilities at length. However, both also call, with some degree of urgency, for intensification of efforts to understand the functioning of managed farm and ranch ecosystems. The selection both of indicators of condition and of performance standards is hampered by ignorance of the causal mechanisms that link farming and ranching practices with resource degradation. Although the general pathways are recognized—the action of heavy machinery in compacting soil, for example—it is usually less clear exactly what the practice contributes to the degradation of the resource. For instance, to continue the example, how much compaction can be tolerated before the soil loses its capacity to absorb rainfall or nutrients?

The question of the adequacy of condition and performance measures is not simply academic. The reauthorization of the Clean Water Act focuses on non-point-source pollution. Agriculture is the nation's largest remaining unregulated source of non-point-source water pollution. Traditional environmental programs focused on agriculture have relied on financial incentives to encourage farmers to address conservation goals, mainly those associated with stopping soil loss. Constraints on public funds make the continuation of hefty incentive payments somewhat problematic and raise the possibility of regulatory fixes, similar to approaches adopted initially to deal with point-source pollution problems.

There are two basic approaches to standard-setting for agriculture. One specifies the production practices or technologies that producers are encouraged or required to use. Such "design standards" are traditionally used in agriculture. The alternative is to establish performance standards. Such standards would set an acceptable level of emissions or some other measurable indicator of environmental quality—for example, nitrate levels in tile drainage, acceptable rates of erosion, or phosphorus levels in surface soils—and allow the producer to determine the best method of meeting those standards. Performance standards leave the producer with the most flexibility to adjust but require more sophisticated scientific and technical capacity to set and monitor. Design standards are easier to set but lock producers into fixed and perhaps more costly and less-effective solutions.

Management of rangelands raises another set of public policy and institutional issues because about half of the nation's rangelands belong to federal or state government. Although private landholders may make management and use decisions with few constraints, public-land managers must often balance competing and conflicting claims advanced in statute or in practice. Consequently, the definition of the condition measure itself is controversial because selection of some indicators over others may imply that less weight is given to one set of values or uses over others. The NRC rangeland report attempts to address that possibility by providing the qualitative basis for a multidimensional index of rangeland health.

If the complementary roles of condition and performance measures are applied to agriculture, the task should be simplified somewhat by the fact that there is usually little uncertainty about what human activity is affecting the ecosystem. Still, it is often difficult to determine the ecological consequences of a given action, and spatial aggregation is a particular difficulty for a site-specific activity such as farming or ranching. Although the scientific basis for performance standards is not well developed, ongoing efforts to manage and alter agricultural systems that pollute provide a wealth of information for designing workable standards. To date, the agricultural community has resisted the development of such standards, preferring voluntary adoption of best-management practices to what appears might be unprecedented mandatory regulation of farming practices. In the long run, though, it will likely be less costly to farming and ranching to work to performance standards rather than to design standards. As experience with other industries has shown, design standards tend to lock technologies in place and discourage development of new ones. As agriculture seeks to take advantage of new information and biotechnologies, impediments to the adoption of innovative management systems and practices could have serious consequences for both agricultural productivity and environmental protection.

REFERENCES

National Research Council. 1993a. Soil and Water Quality: An Agenda for Agriculture. Washington, D.C.: National Academy Press.

National Research Council. 1993b. Restoration of Aquatic Ecosystems. Washington, D.C.: National Academy Press.

National Research Council. 1994. Rangeland Health: New Methods to Classify, Inventory, and Monitor Rangelands. Washington, D.C.: National Academy Press.

Measures of Environmental Performance and
Ecosystem Condition. 1999. Pp. 96–156.
Washington, DC: National Academy Press.

Use of Materials Balances to Estimate Aggregate Waste Generation in the United States

ROBERT U. AYRES AND LESLIE W. AYRES

One can view each industrial sector as a transformation process, where raw material inputs or purchased commodities from upstream sectors and "free goods" from the environment are converted into products for downstream sectors and wastes. This conversion process is subject to the materials-balance constraint not only in the aggregate, but also element by element. In other words, the sum of the weights of all inputs must exactly equal the sum of the weights of all outputs. When both inputs and outputs are known, it is possible to estimate wastes, making due allowance for processes utilizing the free goods (i.e., air, water, topsoil). Of course, in reality, there are often significant uncertainties with regard to either inputs or outputs, or both. These arise from statistical inconsistencies, stock adjustments, imports, and exports. In other work, we have attempted to take all of these into account.

It is important to explain what we mean by macrolevel in this context. In general, large-scale mass flows exceed by many orders of magnitude the flows of the most highly toxic pollutants, including trace elements. In our balancing efforts, we have attempted to construct input-output tables for major process stages. Thus, in the case of agriculture and forest products, we try to balance the flows of carbon, oxygen, hydrogen, and major nutrients. At this level of detail, it is not possible to account precisely for minor flows (e.g., of pesticide residues). Studies accounting for minor flows would have to use different methodology based much more on detailed chemical and metallurgical process data than on economic data.

To avoid unnecessary and distracting biological complications, we treat biomass as a produced good of the agricultural and forestry sectors, even though much of it is arguably free. Unfortunately, this leaves us with the problem of accounting for water, as both an input and output, which cannot be done with great precision. Fortunately, great precision is probably not necessary in this case. Labor and capital inputs, such as machinery, and fuel or electric power for operating the machinery, are not considered explicitly in this paper. However, it should be borne in mind that a considerable fraction of aggregate industrial output is actually capital (and operating) input to other sectors.

Our immediate intention is to classify outputs as either economic commodities or missing mass. In subsequent work, the "missing mass" will be further classified based on the level of waste treatment and final disposal medium (air, water, or soil). This means we need to be quite careful in accounting for the consumption of oxygen (from air) in oxidation processes and for the consumption or production of water in hydration, dehydration, dilution, dissolution, and so on. We selected 1988 as the year of reference for this study because it was the last year for which we had reasonably good international data at the time we began the work. Unfortunately, 1988 was a very atypical year for U.S. agriculture, as we note below.

Agriculture[1]

Inputs to the agriculture sector consist of sunlight, water, carbon dioxide from the air, nitrogen fixation also from the air, topsoil, and some chemicals (e.g., fertilizers and pesticides). Commodity outputs are harvested crops. (Dairy products and meat are considered separately in the next section.) Missing mass, in the aggregate, consists mainly of crop wastes, runoff, evapotranspiration, and oxygen, a by-product of photosynthesis. Other losses include soil erosion, nitrogen (and phosphorus) carried away by water sources, and gaseous emissions.

The production process in agriculture (and also forestry, considered below) can be estimated crudely from the following basic equation of photosynthesis:

$$CO_2 + H_2O + photon \Rightarrow CH_2O + O_2.$$

Plants fix carbon in daylight and release part of it (about half) at night. Water carries nutrients and metabolic products and provides evaporative cooling. There is a rough average proportionality between carbon fixation rate (gross photosynthesis) and evapotranspiration, but there is no fixed relationship between water content and metabolic process; some plant parts are very high in water content, others much less so. In general it seems reasonable to assume that raw biomass contains 50 percent water by weight on average, whereas refined or processed food or feed commodities are considerably drier. In cases where actual data are lacking, we assume 25 percent water content for processed "dry" commodities. Unfortunately, official statistics are not informative on this point.

Raw products of U.S. agriculture include truck crops (fruits, vegetables), tree crops, and field crops (grain, oilseed, hay and alfalfa, sugar beets, sugar cane,

TABLE 1 Agricultural Production in the United States, 1988 (million metric tons)

Commodity	Production Raw	Finished	Exports	Imports	Consumption Raw	Finished
Beef and veal	17.82	10.88	0.31	1.09	11.62	11.20
Lamb and mutton	0.32	0.15	0.00	0.02	0.18	0.18
Pork	9.91	7.11	0.09	0.52	7.51	4.87
Poultry[a]	12.95					6.37
Eggs[b]	4.44					3.86
Dairy products	65.86	64.41				64.41
Subtotal	111.30		2.77	1.26		59.90
Food grains[c]	56.55		44.26		35.40[d]	17.37
Feeds and fodders			11.37			
Feed grains and products[e]	147.06		55.21			
Oilseeds and products[f]	49.16		26.90			
Hay	114.31					
Sugar cane	27.13	2.88		1.21		3.40
Sugar beets	22.51	2.97				3.40
Other field products[g]	5.03		1.57			3.46
Corn syrup						7.64
Subtotal	421.75		139.31			35.27
Vegetables	25.14					22.27
Potatoes	16.17					14.77
Fruits	25.60					11.51
Nuts	0.55					
Fruits, nuts, and vegetables			4.06	6.74		
Coffee, cocoa				1.48		3.40
Subtotal	67.47		4.06	8.21		51.94
Fish	3.26		0.48	3.37	4.77	1.40
Vegetable oils		6.40	1.30	1.35		6.66
SUMMARY						
Animal products	111.30		2.77	1.26		90.87
Field products	421.75		139.31			35.27
Vegetable products	67.47		4.06	8.21		51.94
Fishery products	3.26		0.48	3.37		1.40
Vegetable oils		6.40	1.30	1.35	4.77	6.66
TOTAL	603.71		147.92	14.19		186.15

[a]Poultry conversion at $0.33/lb.
[b]Egg conversion at 0.77 kg/dozen.
[c]Food grains = wheat, rice, rye.
[d]The difference in consumption of raw and finished food grains is used for beer and distilled spirits.
[e]Feed grains and products = corn and sorghum for grain, barley, oats.
[f]Oil seeds and products = soy beans, peanuts, cottonseed, flax seed.
[g]Other field products = dry beans and peas, cotton lint, tobacco.

NOTE: Table does not display stock changes, particularly large in 1988.

SOURCES: Bureau of the Census (1990, 1991) and United States Department of Agriculture (1990, 1991).

potatoes, cotton, tobacco). Harvested output of all field crops, including hay, in 1988 was 421.75 million metric tons (MMT).[2] Truck crops totaled 67.5 MMT. Total weight of harvested crops was 489 MMT (Table 1).[3] It should also be noted that corn plants harvested whole for silage, or "hogged" on the farm, are not included in the grain production figures. This material, which is fed to animals, is classed as "harvested roughage"; it amounted to 68 MMT in 1988. Total biomass harvested by humans and animals including grazing was 885 MMT.

According to one estimate, the average ratio of above-ground crop residues remaining on the land to harvest weight is about 1.5 for cereals (straw), 1.0 for legumes (straw), 0.2 for tubers (tops) and sugar cane (bagasse), and 3.0 for cotton (stalks) (Smil, 1993). On this basis, total residues left above ground in 1988 would have been around 400 MMT. Total above-ground biomass production was about 1,285 MMT. In the United States, most of the crop residues are left on the land; a small fraction, about 20 percent, is burned for fuel or used for other purposes (Smil, 1993). (In China and India, by contrast, as much as two-thirds of crop residue production is burned as fuel in household cooking.)

Biomass is a mixture of cellulosic fiber, carbohydrates, fats, proteins, and water, the latter accounting for about 50 percent by weight. Hence, we estimate that the dry weight of the biomass produced in 1988 was about 642 MMT. For each 100 units of dry output (CH_2O basis), the photosynthetic process equation implies that 146.7 MMT CO_2 (containing 40 units of carbon) were initially extracted from the air, 60 units of water were converted, and 106.7 units of oxygen were returned to the atmosphere. Overall, for 1988, water inputs—not including water required for evapotranspiration—were about 1,027 MMT, and carbon dioxide inputs were about 1,002 MMT. Oxygen produced by the photosynthesis process in agriculture would have been about 685 MMT. The overall flows for U.S. agriculture in 1988 are summarized graphically in Figure 1.

One other air pollutant, methane, is worth mentioning. Most animals have anaerobic organisms in their guts that convert a small amount of the food they consume into methane, typically 1–2 percent on an energy basis. However, this percentage is larger for cattle and sheep, ranging from 5.5 to 7.5 percent, depending on the quality and quantity of feed. Taking these factors into account, Crutzen et al. (1986) have estimated annual methane output of 60 kg per head of cattle and 8 kg per head of sheep. The U.S. cattle population in 1988 was 99.6 million; the sheep population was 10.5 million. Methane emissions from these sources amounted to 0.68 MMT.

It should be noted that the agricultural sector uses large amounts of fertilizers and pesticides. The nitrogen (N) content of ammonia used for fertilizer consumed domestically was 11.2 MMT in 1988, or 76 percent of all the synthetic ammonia produced in the United States that year. Large quantities of urea (about 0.33 MMT), a fertilizer material, are also used for animal feed supplements. Domestic agriculture consumed 33.5 MMT of phosphates containing 10.8 MMT P_2O_5. Many of these substances find their way directly or through animal excreta into surface water and groundwater. Much of the N content of animal feed ends up in

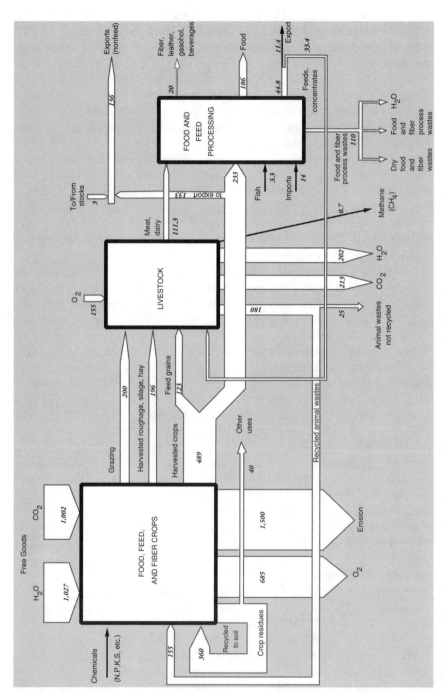

FIGURE 1   Mass flow and carbon, hydrogen, and oxygen balances in U.S. agriculture, 1988 (million metric tons). Calculated by the

urine, either on pastures or at feedlots, resulting in both air and water pollution. We do not have an accurate estimate of the quantities involved, but probably two-thirds of the urine is generated at feeding stations.

Fertilizers and pesticides, direct chemical inputs to agriculture, are not counted explicitly as pollutants, although their use results in pollution. Animal wastes are a major pollution problem, especially in the vicinity of animal feedlots and large-scale poultry producers. Of 100 units of nitrogen in fertilizer, roughly 50 are taken up by harvested crops, of which 47 are subsequently consumed by animals, and 42 of these are eventually excreted as waste (Crutzen, 1976). Most of this waste is generated at feedlots because fertilizer is seldom used on grazing land, and the nitrogen uptake by grazing animals is largely left behind as manure or urine. About 24 units of nitrogen find their way to rivers, lakes, and ground-water, of which 10 units are direct runoff from the soil, 8 come from animal excreta at feeding stations, and 6 from human sewage. Thus, about 18 percent of nitrogen in agricultural fertilizer reappears within a few weeks or months as water-borne pollution, although only 10 percent is due to direct fertilizer loss. Because 11.5 MMT N were used in fertilizer and feed supplements in 1988, this implies an overall waterborne nitrogen-waste flow of 2.76 MMT.

Schlesinger and Hartley (1992, table 4) estimated annual NH_3 emissions per head at 15.5 kg from cattle and horses, 2.4 kg from sheep, 2.35 kg from pigs, and 0.21 kg from poultry. Based on 1988 populations of 99.6 million cattle, 55.5 million pigs, 10.9 million sheep and lambs, and 5.7 billion chickens and turkeys, this totals 2.91 MMT. Fertilizer itself is also a source of ammonia emissions; the emission factors for urea and ammonium sulfate spread on the soil surface are estimated at 0.2 and 0.1, respectively; for other fertilizers—including anhydrous ammonia injected directly into the soil—the emission rate is lower (around 3 percent) (Schlesinger and Hartley, 1992). In 1988, 2.49 MMT N in urea were used as fertilizer in the United States along with 0.340 MMT of nitrogen in ammonium sulfate and 6.84 MMT N from other sources. Altogether, animal metabolism and fertilizer use generated nitrogen emissions of 3.79 MMT (as ammonia). This represented nearly 33 percent of the 11.6 MMT (N content) of ammonia equivalent that was used as fertilizer. Of this, about 8 percent was a direct loss.

The rest of the nitrogen unaccounted for in the applied fertilizer (about 32 percent) is embodied in root and stem material that is not harvested or is harvested directly by animals and remains with the soil (20–25 percent), or is reconverted to nitrogen gas and returned to the atmosphere by denitrifying bacteria in the soil (5–10 percent). For every 16 units of nitrogen emitted as N_2, on average 1 unit is emitted as N_2O, a potent greenhouse gas, but these emissions tend to be episodic. Recently, the use of nitrogenous fertilizer has come to be recognized as one of the major sources of N_2O buildup. Worldwide, an estimated 0.7 MMT N_2O are emitted annually from this source (Schlesinger, 1991). The United States was responsible for roughly one-eighth of worldwide nitrogenous fertilizer use in

1988 and probably a similar proportion of N_2O emissions, together the equivalent of 0.055 MMT N content.

The above estimates do not take into account the relatively small quantities of other chemical elements embodied in the crops, notably nitrogen, phosphorus, and other minerals taken up from the soil or, in the case of nitrogen, fixed by bacteria. It is of interest, however, that the three major chemical elements in dry plant tissue are carbon, hydrogen, and oxygen, which account for 95 percent of the total mass. Nitrogen accounts for another 2 percent, phosphorus for 0.5 percent, potassium for 1 percent, and sulfur for 0.4 percent. These are the major nutrients that are depleted by harvesting and must be replaced by the addition of fertilizers. The remaining 1 percent of plant mass consists of other mineral elements (Table 2) that are readily available from the soil. The flows of nutrients (nitrogen and phosphorus) in U.S. agriculture are summarized in Figure 2.

In 1988, 133 MMT of grain, vegetables, and oilseeds (mostly soya beans) were exported, not including 11.4 MMT of "feeds and fodders," which are from the processing sector. The remainder was consumed directly or indirectly within the United States. Final consumption of all food products (not including beverages) for 1988 was 186 MMT, plus about 20 MMT for grain-based beverages, alcohol, cotton, wool, and other products. Indirect consumption (as animal feed) accounted for most of the difference between gross production and final consumption.

According to the U.S. Department of Agriculture (USDA) (1992), U.S. livestock in 1988 were fed 119.4 MMT of feed grains and 3.7 MMT of food grains

TABLE 2   Chemical Composition of Plants

Element	Percent by Weight
Oxygen	45
Carbon	44
Hydrogen	6
Nitrogen	2
Potassium	1
Calcium	0.6
Phosphorus	0.5
Sulfur	0.4
Magnesium	0.3
Manganese	0.05
Iron	0.02
Chlorine	0.015
Zinc	0.01
Boron	0.005
Copper	0.001
Molybdenum	0.0001
Total	**99.9011**

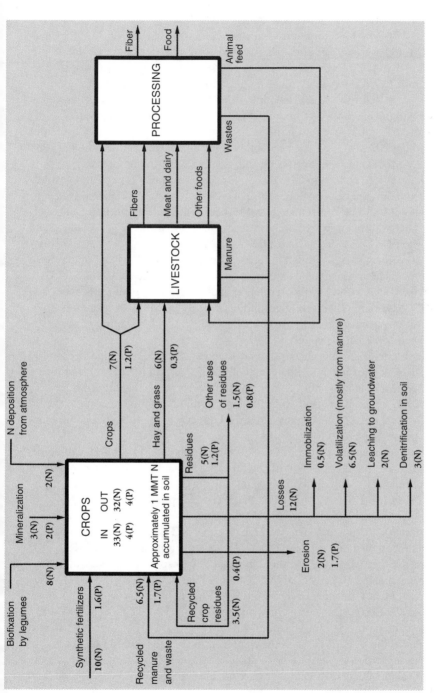

FIGURE 2 Flows of nitrogen (N) and phosphorus (P) in U.S. agriculture, 1988 (million metric tons). Calculated by the authors from various sources, including Smil (1993).

(mostly wheat). Other harvested animal feeds included 123 MMT of hay and alfalfa, 4.76 MMT of sorghum as silage, and approximately 68 MMT of harvested roughage (such as cornstalks) mixed with other feeds, for a total of 319 MMT of harvested feeds. By-products of the food processing industry such as grain mill by-products (e.g., gluten), oilseed meal, brewers and distillers dried grains, meat and fish meal, dried milk, dried beet pulp, and molasses accounted for an additional 33.4 MMT of animal feeds (United States Department of Agriculture, 1992, table 3).[4] Assuming animal intake of pasturage (mainly by cattle) to be about 200 MMT, we can account for total animal feed consumption in 1988 of 552.4 MMT, not including water, salt, urea, or other minor inputs.

Animal feed concentrates in the United States average 79 percent digestibility. From this, we can conclude that 21 percent of the mass of animal feed concentrates (156.3 MMT) fed to dairy cattle, beef cattle in feedlots, hogs, and poultry is passed through immediately in feces. Harvested roughage, or silage and hay (196 MMT), has lower digestibility, probably around 60 percent. This implies 40 percent passes through in feces. Therefore, annual manure output from on-farm and industrial animal feeding operations amounts to about 100 MMT. In addition, USDA (1992) estimates that animal intake from pastures is about 200 MMT. Assuming 60 percent digestibility, roughly 80 MMT of manure is probably left on pastures. This figure could be too low; the digestibility of pasturage may be as low as 40 percent.[5]

Of the total annual manure output of about 180 MMT, it appears that 100 MMT is generated in confinement, and of this, 75 percent (75 MMT) is probably recycled to croplands (Smil, 1993). The remainder of the manure from feedlots (25 MMT, about 50 percent water) is lost to runoff or in other ways. The 80 MMT of manure left on pastures is returned directly to the soil—but not to croplands per se—and does not constitute a waste.

As to the outputs of the livestock sector, a total of 111.3 MMT can be accounted for as the gross weight of animal carcasses and dairy products produced for the market. (See below.) Adjusting for the "excess" water content of raw milk (87 percent water), we assume that the sector produces about 81 MMT of *equivalent* animal products having the same average moisture content as feeds (50 percent). As noted above, feed inputs equal 552 MMT and manure outputs (50 percent moisture basis) are 180 MMT. Simple arithmetic (552 – 180 – 81) reveals that 291 MMT are lost through metabolic (respiration) processes.

An estimated 50 percent of this lost mass (approximately 145 MMT) is carbohydrate (CH_2O) metabolized for energy. This implies that animals consume 155 MMT (1.067×145) of oxygen and produce 213 MMT (1.47×145) CO_2. The oxidation process also generates 87 MMT (0.6×145) of water as vapor in addition to the 115 MMT of water contained in the feed and not otherwise accounted for. Most of this ends up in manure or urine. The water balance is more complicated, of course, because we have not allowed for the water consumed and re-excreted by the animals.

The greatest mass movement from agriculture is the loss of topsoil due to wind and water erosion. A detailed study of topsoil loss in agriculture was carried out by the U.S. Soil Conservation Service in 1982 (Brown and Wolf, 1984). The study found that 44 percent of U.S. cropland was losing topsoil at an unsustainable rate (i.e., faster than the natural rate of soil formation). Topsoil loss in 1982 was estimated at 1,530 MMT. This amount of loss can be assumed to be roughly constant year to year, although optimists believe that the erosion-loss rate is declining as a consequence of increasing use of no-till methods of cultivation. Also, it must be pointed out that eroded material is not necessarily carried out to sea; it may be redeposited on the same field or in the bed of a nearby stream.

To summarize, we estimate overall annual waste from U.S. agriculture as follows: topsoil erosion, 1,500 MMT; undigested and unrecycled feedstuffs (feces) from animals at feeding stations (not including grazing animals on pastures), 25 MMT (50 percent moisture) or 12.5 MMT (dry weight). The latter is mostly undigested cellulose, but includes about 4 percent (0.5 MMT) nitrogen and 1 percent (0.125 MMT) phosphorus. Urine apparently accounts for roughly 42 percent of the total nitrogen content of synthetic fertilizer, or about 4.8 MMT; but this is only the fertilizer contribution. The total must be about three times higher. About a quarter of this (1.2 MMT from fertilizer, 4 MMT total) is volatilized immediately as ammonia; around 2.8 MMT from fertilizer (9 MMT total) ends up in watercourses; the rest goes into groundwater or is recycled back to the land. Ammonia emissions to the atmosphere, direct from fertilizer use, seem to be about 1 MMT N, or 10 percent of inputs, but volatilization losses from manure and urine add another 4 MMT. Other sources (organic decay) add a further contribution. The total for U.S. agriculture is probably around 6.5 MMT per year. Methane emissions to the atmosphere from grazing animals in the United States were apparently about 0.68 MMT (Figure 1). A rough balance for nitrogen and phosphorus is shown in Figure 2.

Food and Feed Processing[6]

The food and feed processing sectors entail a number of activities, including grain and oilseed milling, meat and dairy processing, cotton processing, oil products, sugar production, fermentation industries, baking, confectioneries, and canning and freezing. Unfortunately, USDA does not clearly separate these activities or identify their inputs and outputs.

We estimate inputs to the food processing sector (361 MMT) as the gross agricultural production of harvested crops (489 MMT) less harvested crops fed to animals (123 MMT grains and 114 MMT hay) less exports (excluding exports from stockpiles, 133 MMT), plus animal products (111 MMT), fish (3.3 MMT), and net imports of foodstuffs (14 MMT).

The consumption of domestic food products (flour, prepared cereals, packaged rice, etc.) from all grain mills in 1988 amounted to 17.37 MMT. This does not include grain consumed by the fermentation industries, which produce both alcoholic beverages and fuel alcohol. We estimate that about 19 MMT of grains, mostly corn, were used for fermentation products in 1988. In addition to grain products, 6.4 MMT of vegetable oils and 7.64 MMT of corn syrup were produced by grain and oilseed mills. The fermentation industries, in turn, produced 2.2 MMT of beverage alcohol and 5 MMT of ethanol for fuel, 1 MMT of animal feed concentrates, and an estimated 1 MMT of beverage carbohydrates. To make up the balance, we estimate outputs of 7 MMT CO_2 and 3 MMT of water vapor.

Cotton is a major agricultural product that contributes little to feed and nothing to food. In 1988, the United States produced a net of 9.2 MMT of raw cotton. This was ginned to yield 3.36 MMT of cotton fiber (lint), 5.5 MMT of cottonseed, and 0.27 MMT of linters. Linters are a fibrous material used for felting and cellulosic chemical manufacturing and so are not wasted. The cottonseed was allocated to mills and "other uses," including exports. Mills purchased 4.38 MMT of seeds, of which 3.38 MMT were actually crushed (United States Department of Agriculture, 1992, table 141). The mill product was 0.56 MMT of oil, 1.53 MMT of high-protein cottonseed oilcake, used for animal feed, and 1.29 MMT of milling waste. (The latter is included with overall milling waste in Figure 1.)

Sugar cane weighing 27.13 MMT was reduced to 2.88 MMT of refined cane sugar; similarly, sugar beets weighing 22.51 MMT yielded 2.97 MMT of beet sugar. (About 0.69 MMT of lime was also used in the latter process.) Sugar refining also yielded about 0.59 MMT of molasses (equivalent dry weight), mostly fed to animals. The remainder of the sugar cane waste was mostly cellulosic bagasse. The sugar beet process produces large quantities of pulpy material; about 1 MMT (dry) of this was used for animal feed in 1988 (United States Department of Agriculture, 1992, table 73). The mass disappearance from these two processes alone amounted to about 42.1 MMT. At least half of this mass loss, perhaps as much as 60 percent, or 25 MMT, is water vapor from the various evaporation stages in sugar production. The remaining dry mass is probably burned for energy recovery, although some residues may be discharged into rivers by sewage plants.

Truck crops (vegetables and berries) and tree crops (fruits and nuts) accounted for a harvest weight of 67.5 MMT. Exports took 4.06 MMT and imports added 8.21 MMT, for a total domestic supply of 71.6 MMT. Final consumption, on an as-purchased basis accounted for 51.9 MMT. The difference, 20 MMT, was presumably waste, at both food processing plants and retail stores. We estimate that 60 percent of this mass loss (12 MMT) was evaporative water loss from freeze-drying (e.g., of orange juice) and other processing. The bulk of the food process waste goes into waterways or municipal waste facilities. Some is recovered for other uses, and a small amount may be burned for energy.

Animal products in the United States can be divided into red meats, poultry, and eggs and dairy. The live weight of animals slaughtered for red meat in 1988 was 28.05 MMT. Salable weight of red meat, after processing, was 18.14 MMT, a reduction of roughly 10 MMT, or nearly 36 percent. By-products of meat processing include lard and tallow (about 3.8 MMT),[7] hides (1.02 MMT undressed),[8] dog and cat food, glue, bone meal, blood meal, meat meal, and tankage. The last two items are utilized in animal feed concentrates (2.3 MMT in 1988; United States Department of Agriculture, 1992, table 73). About 2.7 MMT remains unaccounted for. Some of this may be pet food, for which we have no explicit data. We conjecture that most of the missing mass is evaporative water loss in the production of meals and concentrates.

Exports of red meat products in 1988 amounted to 0.4 MMT, and imports, mostly of beef, amounted to 1.63 MMT. Thus, domestic supply of red meat was 19.37 MMT. However, final consumption of meat ("as purchased") was only 16.25 MMT. The difference of 3.16 MMT is waste fat and bone, largely generated by meatcutters in retail shops. This waste ends up ultimately in municipal landfills.

In the case of poultry, live weight was 12.95 MMT in 1988. Dressed carcass weight of poultry for the United States was either 9.5 MMT or 10.07 MMT.[9] This implies a by-product and waste flow at the processing plant of 3.0–3.5 MMT, part of which (2.0–2.5 MMT) is probably recycled as animal feed. The rest, mostly feathers, is dumped or burned. Final consumption of poultry ("as purchased") was 6.37 MMT in 1988. Thus, a further loss of 3.1–3.5 MMT presumably occurred at the retail level. The latter ends up in municipal landfill or in waterways (as biological oxygen demand).

The grain and oilseed milling sector is rather complex. Marketed grain and oilseed milling products consist of vegetable oils (6.4 MMT), flour (17.4 MMT), and corn syrup (7.6 MMT), a total of 31.4 MMT. The imputed output of feed concentrates, by this calculation, is 30.8 MMT (62.2 − 31.4). (We calculated above that 31.6 MMT must have been produced. The match is quite close.)

In summary, we have identified mass losses from the domestic food processing sector as follows: grain and oilseed milling, 9.7 MMT; fermentation, 3.0 MMT; sugar milling, 42.7 MMT; vegetable and fruit canning and freezing, 20.7 MMT; meat and poultry packing, 1.0 MMT (plus about 6.3 MMT in retail shops); dairy processing, 26.1 MMT; egg marketing, 0.6 MMT; and fish packing and retailing, 4.6 MMT. For the sector as a whole, including retail shops, losses total 110 MMT, plus an additional 6.8 MMT of carbon dioxide from fermentation. However, some of this lost mass was not wasted but was converted into animal feeds.

Of the 110 MMT of mass disappearance identified in 1988 (Figure 1), it appears that at least 66.2 MMT consisted of water vapor from evaporative processes in the manufacture of cheese and dried milk products, sugar and corn syrup production, drying and freeze-drying of fruits (such as oranges for juice, prunes, and raisins), and from the production of meat meal and fish meal. An additional 34.5 MMT of mass loss was solid waste of vegetable origin from sugar beets and sugar cane, fruits and vegetables, and grain milling. Some of this consisted of fruit and vegetable skins and stems, nut shells, pits and seeds, inedible leaves, spoilage, and so forth. Usually, these wastes would be generated in quantities too small to be dried and burned efficiently. However, of the dry sugar and grain milling losses (about 27 MMT) a small amount (about 0.1 MMT) was fed to animals. The same was true for part of the missing mass (11.6 MMT) from meat and fish processing.

Even the bulky and "dry" food processing wastes still contain quite a bit of water, probably about 25 percent. If the 27 MMT of bulky combustible waste biomass of vegetable origin is assumed to have been burned for energy recovery, and if the 20 MMT in dry mass is assumed to be chemically similar to cellulose, the CO_2 generated would be around 29.4 MMT, consuming about 21.3 MMT of oxygen and producing about 12 MMT of water vapor in addition to the 7 MMT embodied in the organic material. Of course, the same amount of CO_2 would be generated by natural decay processes, as long as they occur in aerobic conditions.

The material losses that we have identified as likely waste are "dry" in the sense that they do not include the weight of washing, cooking, or process water. They also do not assume a priori mass reduction by combustion of biomass for energy recovery. In this connection, a survey by Science Applications International Corporation (SAIC) (1985) commissioned by the U.S. Environmental Protection Agency (EPA) attempted to identify dry wastes from the industrial sectors. The SAIC estimate of dry weight of wastes from the food and feed processing sector was 6.3 MMT (based on 1976 data). This strongly suggests that combustible wastes were, in fact, mostly burned for mass reduction. A significant fraction of the incombustible organic wastes of animal origin (9.3 MMT) and vegetable origin (7.1 MMT) were actually downstream in the retail sector. Thus, our analysis is consistent with SAIC's results.

Forestry and Wood Products[10]

Wood products are derived from timber tracts, which belong to the forestry sector or are leased from government. As in the case of agriculture, the primary inputs are land, water, and carbon dioxide from the air. Major outputs are timber and oxygen; minor outputs include gums, barks, and maple sap. Several important natural resins and solvents (e.g., turpentine, "naval stores") are derived from gums. Downstream chemical products based on wood distillates include acetone, methyl alcohol, pine oil (pinenes), terpenes, tall oil, and tanning extracts.

Neglecting the minor products, we can construct a rough mass balance for the timber tracts. The data given below imply that the total mass of raw product that was harvested in 1988 was 342 MMT on an air-dried (15 percent moisture) basis, which implies a dry weight of 290 MMT, of which 2 percent consisted of mineral "ash" (see below) and the remainder, 284 MMT, was roughly equivalent to cellulose. The calculated carbon content is thus 114 MMT, which requires an input of 367 MMT of carbon dioxide from the air and 163 MMT of water for photosynthesis. Total oxygen generated was 206 MMT. Assuming the timber had an original moisture content of 48 percent, total water input must have been 545 MMT. Subtracting the water content of the air-dried wood output (51 MMT) from the original water content of the harvested roundwood (268 MMT) implies a water loss from wood dehydration of 217 MMT.

Trees have an estimated carbon-nitrogen-phosphorus ratio of 800:10:1 (Deevey, 1970). Because most of the nutrients are embodied in the bark and foliage, this is an overestimate for harvested wood. However, if it were correct, the nitrogen content of the wood removed from the forests would be on the order of 1.5 MMT, while the phosphorus content would be on the order of 0.15 MMT. As noted below, the phosphorus content of wood ash is about 1 percent, and the ash itself amounts to 2 percent of the total mass of undebarked wood. This implies a total mineral, or ash, content of 6 MMT for the wood harvested and a probable phosphorus content on the order of 0.06 MMT, which is only 40 percent of the loss rate implied by the Deevey C:N:P ratio. Not all of this ash is removed from the forest as some debarking operations are carried out at the logging site.

The total mass of roundwood consumed by U.S. lumber and plywood mills in 1988 was 155.8 MMT. The mass of lumber produced in 1988 was about 49.7 MMT (Ulrich, 1990, table 7). The mass of plywood and veneer produced was 11.1 MMT; hardboard, insulating board, and particleboard amounted to 8.8 MMT. Total processed wood products added up to 69.6 MMT. Allowing for 43.7 MMT of wood chips from lumber mills used for pulping, exports of 6.9 MMT, 11.8 MMT to other uses, and fuel use of 25.8 MMT, there were 4.9 MMT of unutilized waste, equal to about 3 percent of inputs.

The total weight of inputs to pulp was 142.6 MMT, which consisted of just over 105.7 MMT in domestic pulpwood, and 43.7 MMT in chips from lumber mills. U.S. domestic woodpulp production in 1988 was 57.9 MMT (Bureau of the Census, 1991, table 1195).[11] Domestic production of pulp was essentially entirely from domestic resources.[12] Net U.S. exports of woodpulp in 1988 were 0.535 MMT, while chemical uses of dissolving-grade pulp amounted to 1.24 MMT. However, virgin pulp available for domestic paper production was 56.8 MMT. There is a small statistical discrepancy of 0.7 MMT.

Wastepaper collected for recycling in 1988 was reported to weigh 23.8 MMT.[13] To obtain a match between inputs and outputs of the paper sector, we calculate that 17.7 MMT of recycled pulp must have been consumed, of which 4.9 MMT was from internal waste and the rest was postconsumption wastepaper. Allowing for internal recycling, exports, and other uses, 14.2 MMT of wastepaper would have been repulped. About 10 percent (1.4 MMT) of this mass, consisting mostly of inorganic fillers and coatings, would have been lost in the repulping process (United States Congress, Office of Technology Assessment, 1984, figure 24), leaving 12.9 MMT as secondary pulp for paper and paperboard production. Thus, total domestic pulp supply, including internal recycling, was around 74.9 MMT in 1988. Adding 5 MMT of fillers and other chemicals[14] and subtracting 4.9 MMT for internal recycling implies that domestic output of paper and paper products that year was 75.1 MMT.

United Nations data for pulp are subdivided by pulping process (United Nations Statistical Office, 1988). Mechanically produced pulp (5.39 MMT) was virtually entirely used for newsprint, of which domestic production was 5.36 MMT. The other pulp types were, in decreasing order of quantity, sulfate or Kraft pulp (43.53 MMT), semichemical pulp (3.95 MMT), and sulfite pulp (1.415 MMT). Dissolving-grade pulp for chemical use (e.g., in rayon manufacturing) amounted to 1.24 MMT. Again, there is a discrepancy between U.N. and U.S. statistics.

The 142.6 MMT of pulping feeds contained around 21 MMT, or 15 percent, moisture. The 57.9 MMT of pulp output included only 5.8 MMT, or 10 percent, moisture. Overall, then, the apparent mass disappearance between pulpwood and pulp in 1988 was 92.2 MMT, of which 15.2 MMT (21 − 5.8) was presumably water. This would have left 77 MMT bone-dry weight of waste, including ash and chemicals. The bone-dry organic wastes (70 MMT) consisted of lignin, hemicelluloses, and resins. Small amounts of lignin were recovered for use as lignosulfonates; virtually none of the hemicelluloses were recovered for chemical use.[15]

Most of the waste organic material was burned on site to recover energy and chemicals. We do not have data for 1988, but in 1991, the American Forest and Paper Association (AFPA) estimated that 75.3 MMT of the dry waste organic material content of "black liquor" was used as fuel.[16] The energy recycling figure in 1988 was presumably about 2.5 MMT smaller, or 72.8 MMT, based on relative pulp production levels. The AFPA numbers are as close to ours as can reasonably be expected, given the approximations made in our calculation.

All of the pulping processes except the mechanical ones use chemical reagents, notably sodium hydroxide or sulfurous acid, to dissolve the lignin and separate it from the cellulose fibers in the wood. In principle, these chemical reagents are mostly recovered and recycled internally. In practice, of course, recovery is incomplete and some makeup chemicals are required. In fact, makeup requirements and imputed overall losses and wastes are quite considerable.

The sodium sulfate consumed by the U.S. pulp and paper sector in 1988 was 0.48 MMT (Bureau of Mines, 1989, tables 3-5, 12). Similarly, soda ash consumed by the sector was 0.11 MMT. The total elemental sulfur actually consumed was 0.008 MMT. Consumption of lime by the sector was 1.14 MMT. The industry consumed 0.856 MMT of sulfuric acid. Most of these chemicals appear to have been used in pulping.

Other chemical inputs to the pulp and paper industry were used primarily in bleaching. Most virgin chemical pulps for paper are bleached. In 1988, the primary bleaching agents were elemental chlorine (Cl_2), caustic soda, and chlorine dioxide. The latter is manufactured in-house from sodium chlorate, because chlorine dioxide is explosive and too dangerous to ship. In 1988, the paper and pulp industry was the second largest user of chlorine, taking 1.5 MMT, or 14 percent, of total U.S. chlorine output (United States Bureau of Mines, 1989, p. 849). In 1988, the U.S. pulp and paper industry consumed 2.3 MMT of caustic soda, or 24 percent of production (United States Bureau of Mines, 1989, p. 849). It also consumed 0.341 MMT of sodium chloride (salt) (United States Bureau of Mines, 1989, table 18). The other chemical used in large quantities is sodium chlorate. Production of this compound in 1974 was reported to be 0.19 MMT, of which 70 percent was used for pulp bleaching (Lowenheim and Moran, 1975). U.S. production in 1988 was 0.242 MMT, and consumption was probably somewhat higher, thanks to imports. The 70 percent share attributable to pulp bleaching in 1974 is probably a minimum for 1988.[17] On this basis, we estimate that 1988 consumption of sodium chlorate by the pulp sector was at least 0.2 MMT.

The chemicals described above (total weight 7 MMT) were not embodied in the final product and so must be counted as part of the production waste stream. It follows from materials-balance considerations that the annual discharges of chemical wastes from the pulp and paper industry must be roughly equal to the annual inputs, element by element. As it happens, annual "dry" wastes (e.g., sludges) were reported to be 7.7 MMT in the early 1980s (Science Applications International Corporation, 1985). This is much smaller than our estimate of 16 MMT (bone-dry) in losses from primary pulping and bleaching. However, our figures are more plausible because they include not only chemicals (7 MMT) but ash (approximately 1.5 MMT) and some fiber. They are also consistent with mass balance.

Bleaching wastes are mostly 90 percent sodium chloride, but an estimated 10 percent of the chlorine used is chemically bound to lignins and other organic materials in the pulp. This material constitutes a significant part of the process waste. Roughly 6 percent of the mass of the raw pulp is lost during bleaching. The bleaching effluent contains significant quantities of chlorinated organic compounds with very high molecular weights. In fact, 70–95 percent of spent chlorination and alkali extraction liquors have molecular weights greater than 1,000. Such compounds cannot be separated, quantified, or identified by present means. However, measurable traces of dioxins and furans are found among these wastes (Holmbom, 1991).

Kraft-process emissions of greatest environmental concern are noncondensible sulfur-containing gases (hydrogen sulfide, methyl and ethyl mercaptans, dimethyl sulfide, etc.). These are generated at the rate of about 2.5 kg/ton of pulp (International Bank for Reconstruction and Development, 1980). For sulfate pulp in toto, uncontrolled emissions of sulfur-containing gases would have been about 0.1 MMT. The EPA has estimated airborne effluents (excluding CO_2) from the sector to be about 1.15 MMT (United States Environmental Protection Agency, 1991).

A partial list of chemicals used in the paper industry (as opposed to pulping) and embodied in the product includes clay (kaolin) for filling and coating, titanium dioxide for whiteness, and aluminum sulfate (alum) to improve the ink-absorbing quality of printing paper. Allowing for 5 MMT of fillers and other chemicals embodied in paper products, total inputs to U.S. paper mills in 1988 added up to 80 MMT. Paper and pulp products produced domestically from woodpulp amounted to 69.53 MMT according to U.N. data and 75.1 MMT according to U.S. data.[18] The difference is not easily explained.

Within the forest products sector, we can account for about 155 MMT (15 percent moisture basis) of wood and wood wastes burned for fuel in 1988, which is somewhat less than half of the harvested amount. This consisted of 57 MMT harvested for fuelwood (roundwood), 26 MMT wood chips and scrap from lumbering and wood products operations, and 73 MMT of pulping wastes. Of course, the heat energy from the latter two categories of wastes was recovered for use within the industry. Combustion would have required about 74 MMT of oxygen from the air and generated 132 MMT of carbon dioxide and 96 MMT of water vapor.

Wood combustion produces another waste product, wood ash. The combustion of undebarked wood chips and scrap yields 1–2.5 percent ash; debarked wood chips and sawdust yield 0.5–1.4 percent ash (Obernberger, 1994, table 2). We assume that fuelwood averages 2 percent ash (or a total of 1.2 MMT), whereas industrial wood and pulp average 1 percent ash content. Altogether, in addition to wastes already mentioned, we must add 1.8 MMT wood ash. The composite mass flows for the U.S. timber products sector, together with lumber and pulp and paper in 1988, are summarized in Figure 3. Imports and exports are not shown explicitly, although both are significant. The flows are normalized for U.S. consumption of the intermediates, lumber and wood products, and pulp. Thus, the upstream and downstream activities are indicated without reference to actual location. However, there are substantial additional imports of final paper products, but these are not shown.

Roughly 30 percent of the paper and paperboard consumed in the United States was collected for recycling in 1988. The unrecycled fraction of final consumption, amounting to 59.1 MMT, was either burned or disposed of in landfills. Ultimately, all of the unrecycled material is converted to either CO_2 or methane (from anaerobic decay in landfills). If all the decay were aerobic, this would result in 86.6 MMT of CO_2. However, anaerobic decay is probably more prevalent.

Mining (Metal Ores) and Quarrying[19]

There are two main types of waste associated with mining: earth displaced in the process of searching for and removing ore (overburden) and unwanted contaminants (gangue) removed on-site by screening, washing, settling, flotation, centrifuging, and so on. The material shipped to the next stage of processing is, typically, a concentrate that is fed into a downstream process, smelting for metals or combustion for fuels.[20] Smelting wastes are discussed in connection with the

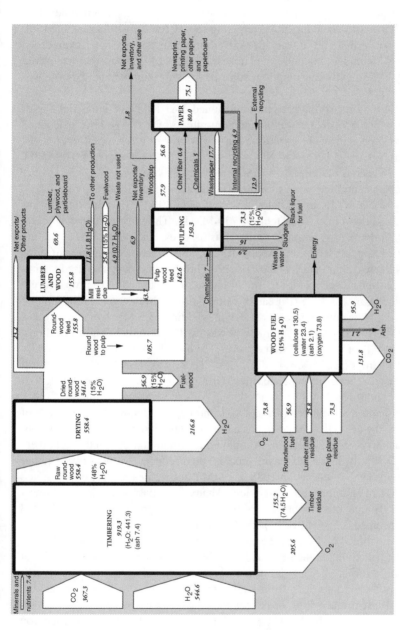

FIGURE 3 Mass flows in the U.S. timber, pulp, and paper industries, 1988 (million metric tons). Calculated by the authors from various sources, including Ulrich (1990), United Nations Statistical Office (1988), Bureau of Mines (1988, 1989), Bureau of the Census (1990), and United States Department of Agriculture (1992).

metallurgical industry, and combustion wastes are discussed separately in connection with fossil fuel combustion.

The Bureau of Mines estimates that mineral exploration and mine development activities (not including those targeting energy fuels) in 1988 generated 190 MMT of waste material, mostly from overburden stripping. Total overburden moved by U.S. mines (excluding coal mines) was 1,241 MMT; overburden moved to supply U.S. metals consumption, adjusted for both imports and exports, was 1,431 MMT.

Mining and quarrying activities consumed 1.87 MMT of industrial explosives, of which 85 percent was ammonium-nitrate based. We have calculated that about 5 percent of the explosive mass was probably converted to N_2O, 14 percent to NO, and 5 percent to NO_2. This implies N_2O emissions of 0.09 MMT and NO_x emissions of about 0.36 MMT from explosives.

Total concentration wastes for metals mined and concentrated domestically, including uranium (discussed below), were about 780 MMT. Adjusting for imports and exports reduces the total to 730 MMT. The most common physical concentration process is froth flotation. It is used, especially, to separate sulfide minerals of copper, lead, zinc, molybdenum, and silver from lighter minerals such as silicates, aluminates, and carbonates. Froth flotation is also used to concentrate phosphate rock and, to a minor extent, to clean coal. The most recent data available are for 1985 (Bureau of Mines, 1987). In that year, 380 MMT of mineral ores were concentrated by flotation, yielding 71.5 MMT of concentrates and 309 MMT of dry equivalent mineral wastes. There were 947 billion gallons, or 3,580 MMT, of water used in the process. Wastes from flotation are generally disposed of in ponds, mostly in dry areas. There was 0.63 MMT of chemical reagents used in concentration activities. Grinding mills required 8 billion kW of electricity and 0.134 MMT of iron rods and balls to break up the lumps of ore.

Aluminum and iron ores are not concentrated by flotation. In the case of aluminum, the ore is bauxite, which is a relatively pure mineral that already contains about 30 percent aluminum. Aluminum is further concentrated to relatively pure Al_2O_3 by the so-called Bayer process, then reduced by electrolysis rather than carbothermic smelting. In 1988, 8.2 MMT of bauxite ore were concentrated to 4.6 MMT of alumina (and some calcined bauxite) in the United States. Almost all of the bauxite was imported. Primary aluminum production in the United States consumed 3.2 MMT of imported alumina.

Iron ore mined in the United States is concentrated for blast furnaces by two processes: pelletizing at the mine and sintering. The latter process is carried out near the blast furnace because it utilizes significant quantities of iron-rich reverts, such as mill scale and dust, from later stages in the iron and steel production process. Blast furnace feed, or concentrates, average 63 percent iron, whereas domestic iron ore is only about 20 percent iron. In 1988, 197 MMT of crude iron ore were concentrated into 57.5 MMT of furnace feed, leaving 140 MMT of wastes.

Large amounts of nonmetallic minerals are mined or quarried domestically. By weight, stone (1,150 MMT, including limestone) and sand and gravel (863 MMT) top the list of these minerals. Imports and exports are comparatively small. Unlike the case of metals, overburden wastes are small in relation to production (except for clay, for which waste amounts to 35.7 MMT). Domestic concentration wastes are also negligible, except for phosphates (179 MMT), potash (8.9 MMT), and soda ash from trona (7.0 MMT).

Phosphate rock mining and processing are extremely important activities, because phosphate fertilizers are absolutely essential for modern agriculture. Unfortunately, the ore is not of very high grade and is contaminated, especially with cadmium, fluorine, and uranium. In the United States, 451.8 MMT of raw materials were handled to produce 224.1 MMT of crude phosphate rock in 1988. The difference was presumably overburden, which was mostly left in previously mined areas. The crude ore was concentrated, mainly by flotation, to 45.4 MMT of concentrated fluorapatite mineral—roughly $(CaF) \cdot Ca_4(PO_4)_3$—which was, in turn, treated by sulfuric acid to yield fertilizer-grade phosphoric acid (13.8 MMT phosphorus pentoxide [P_2O_5]). This refining operation is considered to be part of the chemical industry and is not discussed further here.

Uranium mining in the United States produced about 15 MMT of ore in 1980. This was reduced, mostly by flotation, to 19,500 metric tons (MT) of U_3O_8 concentrate, or yellow cake, which yielded 4,740 MT of refined uranium oxide (nuclear fuel). About 3,200 MT of ore are needed to produce 1 MT of concentrated UO_2 pellets (LeBel, 1982, table 6.1). Uranium production has been declining sharply; production in 1991 was 0.58 MMT of ore and 1,150 MT of yellow cake, a decrease of 96 percent from 1980. Uranium mining added 15 MMT to the 1980 figure for concentration waste, but due to declining demand, waste amounted to about 1 MMT in 1988. (We do not have an exact figure for that year.)

Mine wastes from metallic and nonmetallic mineral production within the United States in 1988 can be summarized as follows. For metal ores, overburden wastes were 1,192 MMT and concentration wastes were 784 MMT, if alumina and crude phosphate rock processing is included, or about 965 MMT and 600 MMT, respectively, if it is not (alumina and phosphate rock are considered products of the chemical industry). The metals system is discussed further below and summarized diagrammatically in connection with metals smelting and refining.

For nonmetallic minerals excluding coal, overburden wastes were 47 MMT and concentration wastes were 15.9 MMT (excluding phosphates). The nonmetallic minerals that are ultimately transformed into inorganic chemicals (phosphates, potash, and soda ash) are discussed further below in connection with chemicals.

The total of overburden wastes for metals and nonmetallic minerals amounted to 1,239 MMT; total concentration wastes were 800 MMT, essentially dry weight, including alumina and phosphates (Tables 3 and 4). Total mineral mining wastes in the United States are actually about 2,050 MMT, excluding water used for

TABLE 3 Production and Waste Allocation for Primary U.S. Metal Production, 1988 (1,000 metric tons)

Metal	U.S. Domestic Mine Production — Total Material Handled (A)	Ore Treated or Sold (B)	Overburden (A − B)	U.S. Domestic Concentrate Production — Production (metal content)	Production (gross weight)[a] (C)	Concentration Wastes (B − C)	U.S. Primary Metal Production (domestic and foreign ores) — Concentrate Consumption (gross weight)[a] (D)	Primary Production (total) (E)	Smelting/ Refining Losses (D − E)
Bauxite/ Aluminum	8,246[b]	1,107[b] 8,970[c]	7,140		4,575	4,395	7,730	3,944	3,786
Copper	523,446	218,631	304,814	1,341	5,364	213,267	5,794	1,406	4,388
Gold	536,146	117,934	418,212	0.201	0.201	117,934	392	0.138	392
Lead	9,707	6,450	3,257	385	481	5,969	490	392	98
Molybdenum	127,006[d]	72,212	54,794	43	172	72,040	103	26	77
Platinum group	34,189	11,396[e]	22,793	0.005	0.005	11,396	0.0003	0.0003	
Silver	48,444	15,876	32,568	1.661	1.661	15,874	2	1.718	
Zinc	21,149	9,106[e]	12,043	244	432	8,674	429	241	188
Uranium oxide[f]	22,000	15,200	6,800		20	15,180	20	5	15
Nonferrous total	1,330,333	467,912	862,420		11,046	460,334	14,960	6,016	8,944
Iron total	300,278	197,766	102,512		57,515	140,251	83,694	49,242	34,452
TOTAL	1,630,611	665,678	964,932		68,561	600,581	98,654	55,258	43,396

[a]Where direct figures for gross weight of concentrate were unavailable, they were calculated by applying reasonable concentration ratios to the metal content.
[b]Bauxite at U.S. mines. Included in total of "ore treated."
[c]Crude bauxite ore, dry equivalent. Includes net imports. Not included in "ore treated" total.
[d]Assumes 3:1 ratio material handled to ore, as with gold/silver.
[e]Zinc data have been subtracted from "other" in 1989 Minerals Yearbook to construct approximate platinum data.
[f]1980 data.

SOURCE: Bureau of Mines (1988, 1989).

TABLE 4 Production and Waste Allocation for U.S. Industrial Mineral Production from Domestic and Foreign Ores, 1988 (1,000 metric tons)

Mineral	Total Material Handled (A)	Ore Treated or Sold (B)	Domestic Production (C)	Exports (D)	Imports (E)	Apparent Consumption (C − D + E)	Overburden Loss (A − B)	Concentration Losses (B − C)
Abrasives, natural	232	156	156			156	76	0
Barite	404	404	404	0.205	1,207	1,611	0	0
Clays	83,370	44,633	44,515	3,535	33	41,013	38,737	118
Diatomite	3,420	695	629	147	0	482	2,725	66
Feldspar	649	649	649	12	287	924	0	0
Gypsum	18,325	14,869	14,869	246	8,782	23,405	3,456	0
Mica	130	130	130	6	12	136	0	0
Perlite	586	586	585	0	0	585	0	1
Phosphorus	451,778	224,075	45,389	0	0	45,389	227,703	178,686
Potassium salts	12,247	11,884	2,999	579	6,964	9,384	363	8,885
Pumice	423	374	353	1	28	380	49	21
Salt	34,470	34,470	34,470	802	4,966	38,634	0	0
Sand and gravel	863,640	863,640	863,531	1,837	357	862,051	0	109
Soda ash	15,728	15,728	8,738	2,238	257	6,757	0	6,990
Stone (estimate)	1,151,000	1,150,000	1,148,533	3,304	3,300	1,148,529	1,000	1,467
Talc, soapstone, pyrophilite	1,179	1,234	1,234	382	80	932	−54	0
Vermiculite	3,393	1,769	275	18	32	289	1,624	1,495
TOTAL	2,640,974	2,365,295	2,167,459	13,108	26,305	2,180,656	275,679	197,836

SOURCE: Bureau of Mines (1988, 1989).

flotation. It is interesting to contrast this number with the only other published estimate we could find, for 1985, of 1,400 MMT (Science Applications International Corporation, 1985). The SAIC figure is far too low because it supposedly included the coal mining sector, which we discuss below.

Mineral Fuels: Coal Mining and Oil and Gas Drilling[21]

Coal mining is the largest single source of waste materials. For every short ton of coal moved, 6.5 tons of mostly overburden wastes are produced (Anonymous, U.S. Department of Energy, personal communication, 1994). Because national soft coal production in 1988 was 862.1 MMT (weight at the mine), total materials handled in coal mining, exclusive of the coal itself, were on the order of 5,600 MMT. This is more than three times the amount of topsoil lost by erosion. Coal mining is also a source of methane; methane is trapped in the coal seams and released when the coal is pulverized. Later, we consider methane production from all fossil-fuel-related activities.

Some utility coal is washed to remove pyrites and ash, resulting in a significant further production of waste. In 1975, about 41 percent of soft coal was cleaned, resulting in 16 tons of coal refuse for every 100 tons of coal washed, for an 84 percent yield.[22] By 1988, more low-sulfur western coal was mined and only about 30 percent of midwestern coal was cleaned. In 1988, therefore, we assume that 4.8 tons (0.3 × 16) of sulfurous refuse were produced at the mine per 100 tons of coal mined. Given that 862 MMT were shipped, approximately 900 MMT must have been mined, generating beneficiation wastes of about 47 MMT, give or take 10 MMT. Sulfurous refuse is a significant cause of acid mine drainage. The 1983 Census of Manufactures reported that coal mines discharged 470 MMT of water, including washing water (Bureau of the Census, 1983). Combustion emissions are discussed below.

In 1988, 86.2 MMT of U.S. coal were exported and 1.94 MMT were imported. Roughly 42 MMT of coal went to coke ovens, producing about 30 MMT of finished coke, 7.6 MMT of coke oven gas, 3.9 MMT of tar and breeze, and small quantities of other by-products, including ammonium sulfate. There were also minor fugitive emissions, mainly from the coke quenching process. The mass flows and wastes in the coal system are shown in Figure 4.

Petroleum and natural gas production involve relatively little waste, except water. During 1988, the U.S. oil and gas industry drilled 25,000 wells encompassing 124 million feet of holes. Assuming 6-inch pipe for the holes, and 5.5 liters of material removed per linear foot, this drilling would have generated 682 million liters total. A liter of water weighs 1 kg, and we assume drilling wastes displaced by pipe have an average specific gravity of 3, so the weight of displaced materials was about 2 MMT. Another 4–5 MMT of material was removed and displaced by the drilling mud (allowing for water content). We therefore estimate a total of 6–7 MMT of drilling wastes.

119

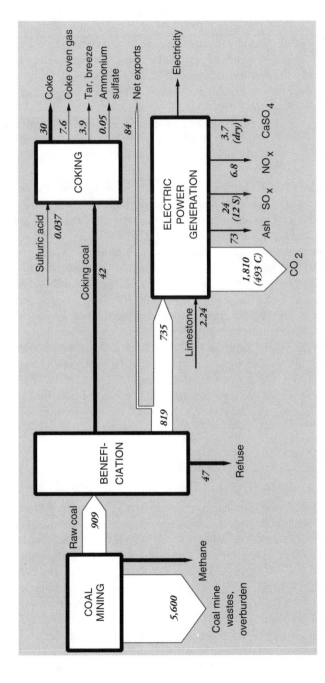

FIGURE 4 Mass flows in the U.S. coal sector, 1988 (million metric tons). Calculated by the authors from various sources, including Bureau of Mines (1988, 1989).

Drilling muds constitute a much larger source of waste. Drilling muds are, on average, 86 percent water, much of which is taken from the wells themselves, 3 percent oil, 2 percent polymers, and 9 percent other materials (United States Congress, Office of Technology Assessment, 1992). The latter include clay, barite, and chrome lignin sulfonates. It should be noted that petroleum drilling accounts for virtually the entire national consumption of the mineral barite, or 1.4 MT. Consumption of clays, notably bentonite and fullers' earth, accounted for 1 MMT. On this basis, drilling muds must have weighed at least 25 MMT. EPA has estimated that drilling fluids used in 1985 weighed 57 MMT (United States Congress, Office of Technology Assessment, 1992). Additional "associated wastes" amounted to 2 MMT, most of which are stored in ponds, where the water gradually evaporates.

EPA estimated that 3.7 billion tons of "produced waters" were generated by drilling activities in 1985 (United States Congress, Office of Technology Assessment, 1992). About 62 percent of this water was reinjected in oil and gas recovery operations, leaving 1.4 billion tons, or 1,270 MMT. Produced waters are usually saline and therefore constitute a disposal problem of some magnitude. (Most of the water is injected into wells.) The excess of produced waters would help to reconcile the water data for the mining sector as a whole. The 1983 Census of Manufactures reported that about 700 MMT of water were used for cooling in the gas-fractionation process (Burea of the Census, 1983).

Natural-gas distribution by pipeline is a source of methane leakage to the atmosphere. In the United States in 1988, it is estimated that 17.15 MMT, equal to 3.5 percent of total production and 4.6 percent of the total quantity transported, disappeared from the system (International Energy Agency, 1991). We estimate that half of this was used to drive compressors whereas the other half, approximately 8 MMT, leaked into the atmosphere. Further losses occur in local distribution.

Petroleum Refining[23]

Total output of crude oil in the United States in 1988 was 402.6 MMT. Exports were 0.6 MMT, and imports were 269.05 MMT, for a total domestic crude oil supply of 671.0 MMT. Reported inputs of crude oil to domestic oil refineries were 680.687 MMT, leaving a discrepancy of 9.1 MMT. In addition, refineries consumed 17.15 MMT of natural gas as fuel, 16.230 MMT of natural gas liquids, and 22.585 MMT of intermediate feedstocks. Of the latter, 16.166 MMT were imported and 6.319 MMT were internal transfers of "gasoil" to be upgraded. Subtracting the gasoil amount from both inputs and outputs leaves a total of 730.23 MMT of net inputs (Table 5).

Crude oil is desalted before refining. Water pollution from this process contains emulsified oil as well as salts, ammonia, sulfides, and phenols. Also, this process involves considerable water use. Refinery products include noncon-

TABLE 5 U.S. Energy Statistics, 1988 (million metric tons)

Commodity	Production Raw	Production Finished[a]	Exports Raw	Exports Finished	Imports Raw	Imports Finished	Consumption Raw	Consumption Finished
Refinery Inputs								
Crude oil	402.585		0.634		269.053		680.687	
Feedstocks						16.864	22.585	
Natural gas liquids		51.325		1.617		7.108	16.230	
Refinery Products								
Liquid petroleum gas		15.906				19.351		56.272
Motor gas		297.439		1.069		0.020		315.690
Aviation gas		1.020						1.074
Jet fuel		63.935		1.362		3.802		66.893
Kerosene		4.347		0.054		0.173		4.689
Diesel		143.258		3.419		12.943		146.315
Residual fuel oil		56.789		11.044		32.818		61.429
Naphtha		6.170		0.479		3.216		8.864
Petroleum coke		36.131		15.344		0.110		20.709
Other		54.073		2.626		5.865		57.072
Total Petroleum	402.585	679.068	0.634	35.397	269.053	78.298	719.502	739.007
Coal and coke	862.066	29.397	86.203	0.992	1.936	2.439	801.647	30.844
Natural gas (18.02 kg/TJ)	424.990	335.028				2.870		302.100
Total other fuel	1,287.056	364.425	86.203	0.992	1.936	5.310	801.647	332.940
TOTAL	1,689.640	1,043.493	86.837	36.389	270.989	83.610	1,521.149	1,071.947

[a]Finished values may exceed raw values because "finished" refers to refinery products whose inputs may have been imported.

SOURCE: International Energy Agency (1991).

densible refinery gases amounting to 34.072 MMT and salable products weighing 679.068 MMT, or 672.749 MMT after subtracting the 6.319 MMT of gasoil that was internally recycled. Petrochemicals are also derived to some extent from "light ends," dissolved volatiles that result from petroleum refining.[24] Crude oil contains relatively little of these materials. A typical U.S. refinery using Texas or Louisiana crude oil might yield 1.3 percent light ends (as refinery off-gas) by volume from the initial distillation (Gaines and Wolsky, 1981). However, subsequent refining processes such as catalytic cracking, catalytic reforming, and delayed coking also yield large quantities of light ends. Most of these by-product volatiles are used internally within the refinery complex. The C_4 gases (butylene, isobutane, and n-butane) are mostly alkylated or blended directly into gasoline. The C_3 gases (propane, propylene) are collected and liquified under pressure for use as domestic fuel (i.e., liquid propane gas). The mixed C_1 and C_2 gases (methane, ethane, and ethylene) are mainly used as fuel for steam generation to provide heat energy for the refinery itself. Large amounts of hydrogen-rich off-gases are produced in refineries, but these are mostly used for hydrotreatment of naphtha or for hydroforming within the refinery. Similarly, catalytic reformate, derived from naphtha, is the major source of aromatic feedstocks, known as BTX (benzene-toluene-xylene), although most of this material is blended into gasoline to increase the octane number.

Nevertheless, light ends also constitute a source of aliphatic petrochemical feedstocks. In brief, light alkane feedstocks such as ethane (C_2H_6), propane (C_3H_8), and butanes (C_4H_{10})—along with some naphtha and heavy gas oil—are dehydrogenated in a pyrolysis furnace within the refinery complex to yield ethylene (C_2H_4), propylene (C_3H_6), butadiene (C_4H_6), butene, butylene (C_4H_8), and other C_4 olefins. In 1988, crude petroleum-based feedstocks consisted of liquified petroleum gas (22.459 MMT), naphtha (8.864 MMT), and light ends (1.12 MMT). (See section on organic chemicals, below.)

In summary, apparent mass losses during refining in 1988 amounted to 55.58 MMT (728.33 − 672.75), or 7.6 percent of input mass. In other words, the efficiency of mass conversion was 92.4 percent. Presumably, virtually all of the missing mass consists of carbon dioxide, carbon monoxide, or hydrocarbons, including fugitive volatile organic compounds (VOCs). The EPA estimated airborne effluents from the sector at 2 MMT (United States Environmental Protection Agency, 1991); we assume this figure does not include carbon dioxide. This is consistent with materials-balance arguments. VOCs from petroleum refining include significant quantities of BTX and other aromatics, many of which are carcinogenic.[25]

Crude oil contains small quantities, on the order of 0.1 percent depending on its origin, of sulfur and mineral ash. For example, Venezuelan oil is particularly high in sulfur. The petroleum refining industry recovers sulfur from crude oil and produces sulfuric acid as a by-product (2.4 MMT H_2SO_4 or 0.786 MMT S). Most of this sulfuric acid is used within the refinery for bleaching. However, over half

(1.3 MMT) of the spent acid is recovered and sold. Most of the ash in crude oil probably remains with the refinery solid wastes and sludges. Assuming the crude oil contains 0.1 percent ash, there would be some 0.7 MMT of solid waste. The spent sulfuric acid (2.1 MMT) is presumably neutralized, either by reaction with some of the alkaline minerals (NaO, KO, MgO, CaO) in the crude oil ash, or by added lime.

Refineries also use materials purchased from other sectors in the refining process, including salt (0.72 MMT) and clays (0.122 MMT), which subsequently reappear mostly in solid or liquid wastes. Consumption of salt is reported, but its use is unclear; it may be a precursor to in-house caustic soda production (Gaines and Wolsky, 1981).

Mass flows in the petroleum, gas, and refinery sectors are shown in Figure 5. As noted above, the mass loss in refineries amounted to 55.6 MMT in 1988, or 7.6 percent of the mass of hydrocarbon inputs. This loss is partly fugitive emissions (VOCs), estimated to be 2 MMT, but most of these emissions are flared or recovered for refinery heat and power. Assuming that the missing mass consists mostly of molecules of the form C_NH_{2N+2}, we can safely predict carbon content of between 80 and 85 percent, or about 45 MMT. This corresponds to an atmospheric oxygen intake of 120 MMT, eventual CO_2 output of roughly 165 MMT, and water-vapor output of 76.5 MMT.

Dry wastes from the petroleum refining sector were about 1.25 MMT in 1981 (Science Applications International Corporation, 1985). Given the fact that neutralized sulfuric acid wastes alone would account for considerably more than this, we estimate the solid wastes from refineries to be 3–4 MMT (i.e., three times more than the SAIC estimate). EPA's estimate of total nonhazardous waste produced by the sector in 1985 was 150 MMT. Because the missing mass in the sector was only 55.6 MMT, including purchased inputs, and most of this must have been combustion products and VOCs, it is clear that most of the waste mass counted by EPA must have been water.

Chemicals

The chemical industry is far too complex to describe in adequate detail in the space available here. Our discussion is inevitably somewhat superficial. The major distinction is between inorganic and organic chemicals. The latter are derived mostly from natural gas, natural gas liquids, or petrochemical feedstocks.

Inorganic Chemicals[26]

Inorganic chemicals are derived either from nonmetallic minerals such as sulfur, phosphates, potash, soda ash, and salt or from the atmosphere. A few inorganics are derived from metal ores or metals. A summary diagram for phosphates, potash, and soda ash, beginning with extraction and beneficiation, is given

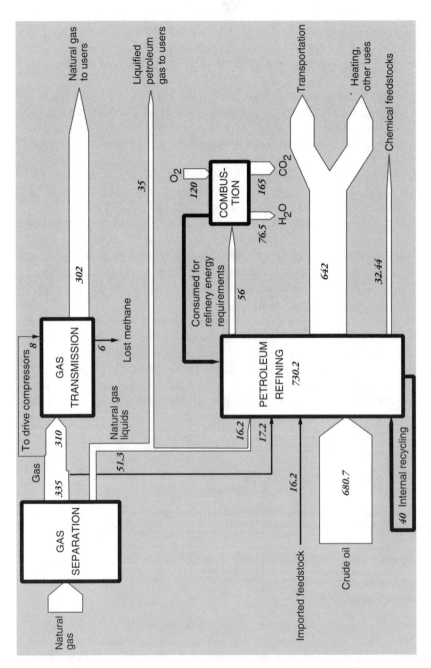

FIGURE 5 Mass flows in the U.S. oil and gas sectors, 1988 (million metric tons). Calculated by the authors from various sources, including International Energy Agency (1991), Subak et al. (1992), and United States Environmental Protection Agency (1988, 1991).

in Figure 6. The most important inorganic chemicals other than the three noted above are ammonia, sulfur (sulfuric acid), and salt (chlorine and sodium hydroxide) (Figure 7). The major groupings are discussed in the following sections. Others are mostly derived from these. For example, nitric acid and urea are both made from ammonia.

Domestic production of ammonia in 1988 was 12.544 MMT (N content). Net imports (imports minus exports) plus stock changes increased apparent domestic consumption of ammonia to 14.745 MMT (N). In addition, there were significant imports and exports of nitrogen-containing chemicals. The major net import items were urea (0.483 MMT N) and ammonium nitrate (0.091 MMT N), while major net export items were ammonium phosphates (1.150 MMT N) and ammonium sulfate (0.155 MMT N). In all, fertilizers accounted for nearly 80 percent of the supply of fixed nitrogen (Figure 7).

The nitrogen content of monomers embodied in plastics and resins in 1988 added up to 0.669 MMT. Nitrate and nitro-explosives, excluding amines, accounted for about 0.777 MMT (N). Urea fed to animals accounted for 1.55 MMT (N); unspecified uses of nitric acid, including phosphate rock processing and steel pickling, accounted for 0.135 MMT (N). Other identifiable final uses, including dyes, rubber chemicals, herbicides and pesticides, and sodium cyanide used in the gold mining industry, added to 0.153 MMT (N). Process losses are probably at least 2 percent. (Nitrogen wastes are less than they might otherwise be, however, because ammonia-bearing waste streams are easily neutralized by sulfuric acid to produce a useful by-product, ammonium sulfate fertilizer.) We estimate that process losses altogether account for 0.3–0.35 MMT (N). There is some possibility of undercounting of the use of nitrogen in mixed fertilizers where published data seem to be spotty. However, lacking further information, we assume the remaining "missing" nitrogen (about 0.4 MMT) is allocated mostly to household cleaning agents and other consumer products. In summary, we can account for about 0.737 MMT of fixed nitrogen embodied in products of the synthetic organic chemicals sector and on the order of 0.16 MMT of nitrogenous losses associated with organic synthesis, for a total of 0.9 MMT.

In terms of environmental pollution, the 2 percent loss rate suggested above is insignificant compared with dissipative uses of nitrogenous chemicals. Apart from fertilizers and animal feeds, these include industrial explosives, pesticides and herbicides, dyes, surfactants, flotation agents, rubber accelerators, plasticizers, gas conditioning agents, and so on. In fact, except for plastics and resins (and plasticizers), it is safe to assume that virtually all nitrogenous chemicals are soon dissipated in normal use, but mainly by other sectors or final consumers. In the case of plastics and fibers, the dissipation is merely slower.

Sulfuric acid (37.7 MMT in 1988) is derived from elemental sulfur; 11.584 MMT of sulfur were produced in 1988, most of which (10.3 MMT) was used to produce sulfuric acid, which is the starting point for most sulfur-based chemicals (Figure 7). Elemental sulfur is also recovered from natural-gas processors and

126

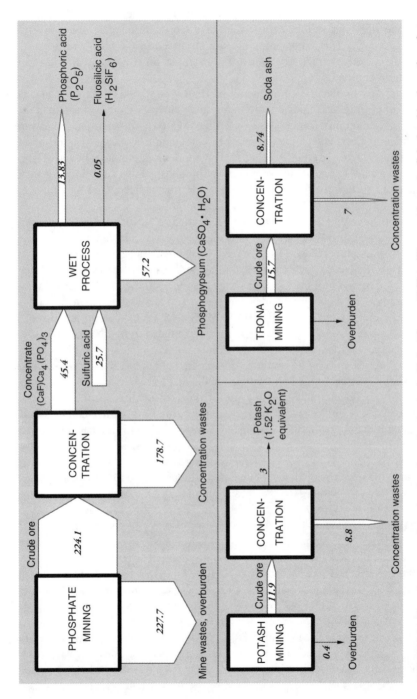

FIGURE 6 Phosphates, potash, and soda ash production in the United States, 1988 (million metric tons). Calculated by the authors from various sources, including Bureau of Mines (1989).

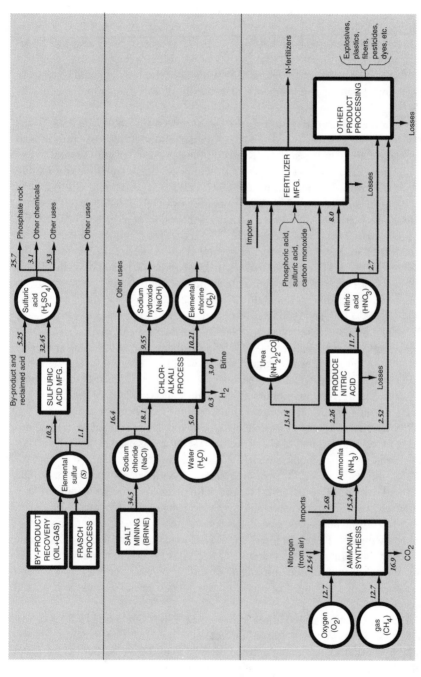

FIGURE 7   Mass flows associated with production of sulfur compounds, salts, and ammonia in the United States, 1988 (million metric tons). Calculated by the authors from various sources.

petroleum refineries. In the latter case, most (0.958 MMT) of the sulfur was used on site, but almost half of the spent acid (0.43 MMT S) from refineries was later reclaimed. Sulfuric acid was recovered from copper, zinc, and lead smelters (1.125 MMT S), but much of that (0.543 MMT) was used in the acid-leach process of mining operations. In the case of copper mines, copper sulfate is recovered from the leach piles and recycled, but much of the leaching acid remains in the ore heaps, where it presumably reacts with other minerals and remains as insoluble sulfates.

By far the largest use of sulfuric acid is for processing phosphate rock (8.404 MMT S). This use was noted above. Another important use of sulfuric acid (0.288 MMT S) is in the sulfate (Kraft) pulping process, also discussed above. The pickling process used to clean rolled steel prior to galvanizing or tin-plating used 0.074 MMT S as acid, of which 0.022 MMT was recovered. Sulfuric acid containing an additional 0.024 MMT S was used by other metallurgical processes, mainly metal plating. Automotive batteries accounted for a further 0.051 MMT, of which 0.036 MMT was reclaimed. The above uses plus exports add up to 11.162 MMT of embodied sulfur, or greater than 90 percent of the total. The remainder, 1.172 MMT of embodied sulfur, either is used elsewhere in the chemical industry or is used for unidentified nonchemical purposes.

Excluding phosphates and sulfuric acid itself, about 5.04 MMT of sulfuric acid (1.645 MMT S) and 0.684 MMT of elemental sulfur were consumed by the nonphosphate chemical industry in 1988. Industrial inorganic chemicals, including pigments, consumed about 0.9 MMT S, mostly as acid. Of the sulfur used in chemicals, 0.566 MMT was eventually converted into ammonium sulfate fertilizer, mostly as a by-product of other chemical processes that use sulfuric acid (e.g., caprolactam, a nylon monomer, and hydrogen cyanide). Of the rest, 0.170 MMT was embodied in aluminum sulfate, mainly for the paper industry; 0.185 MMT S was in the form of by-product sodium sulfate, consumed in pulp manufacture. Apparently, 0.460 MMT of elemental sulfur was used in "other" agricultural chemicals. The organic side of the chemical industry consumed at least 0.7 MMT S, mostly as acid. Of this, drugs and pesticides accounted for only 0.02 MMT and detergents for 0.06 MMT. The major uses were for the manufacture of organic chemical intermediates and synthetic rubbers and plastics. However, except for drugs, pesticides, and detergents, virtually no sulfur is embodied in organic chemicals. Hence, at least 0.6 MMT S was consumed and lost in organic processing. In addition, there was 0.824 MMT S as sulfuric acid and 0.297 MMT elemental sulfur in the "unidentified and export" category. For reasons discussed below, we believe that most of this was also consumed in the organic chemical industry.

These waste flows were in several chemical forms, including H_2S, SO_2, ammonium bisulfate, calcium sulfite, and calcium sulfate. (To mention one example, hydrofluoric acid manufactured in 1988 would have accounted for about 0.24 MMT S and generated calcium sulfate waste.)

Phosphate rock is the only source of phosphorus chemicals, including fertilizers. The starting point is fertilizer-grade phosphoric acid (13.833 MMT P_2O_5 content).[27] Of 1988 production, exports—mostly ammonium phosphates—accounted for 2.608 MMT, leaving 10.549 MMT for domestic consumption. Most of the latter, 9.329 MMT, or 88.4 percent, was converted into "wet process" phosphoric acid (H_3PO_4). Elemental phosphorus production in the United States in 1988 was 0.32 MMT (0.73 MMT P_2O_5 equivalent), of which 85 percent was reconverted to pure furnace-grade phosphoric acid for chemical manufacturing. Some of this is used to make triple superphosphate fertilizer, but about 40 percent was used to manufacture sodium tripolyphosphate (STPP, $Na_5P_3O_{10}$), a detergent builder. Production of this chemical in 1988 was 0.497 MMT, with a P_2O_5 equivalent of 0.309 MMT. The detergent industry has been shifting to an alternative, tetrasodium pyrophosphate, which contains less phosphorus. Some phosphoric acid is used as a flavoring agent in the food and soft drink industry.

A minor but growing use of phosphorus is in the manufacture of lubricating-oil additives such as zinc dithiophosphate. This use accounted for 0.015 MMT of phosphorus metal in 1974; we estimate 0.02 MMT in 1988, corresponding to about 0.05 MMT P_2O_5 equivalent.

The starting point for organic phosphate synthesis is phosphorus trichloride (PCl_3). Production figures are not published, but assuming 1 percent of chlorine output goes to PCl_3 production (see discussion of chlorine below), we conclude that about 0.03 MMT of phosphorus metal, or 0.07 MMT P_2O_5, would have been required. The trichloride is later converted to phosphorus oxychloride ($POCl_3$) by direct reaction with chlorine and phosphorus pentoxide P_2O_5. The oxychloride, in turn, is the basis of organic phosphate esters that now have many uses. The most important of these uses is in the manufacture of the plasticizer tricresyl phosphate. Phosphate esters are also used as flame retardants and fire-resistant hydraulic fluids. The United States used 0.043 MMT (P_2O_5) of phosphate esters in 1990 (International Trade Commission, 1991). Usage in 1988 was probably similar.

All of the uses mentioned above are inherently dissipative. No more than 0.2 MMT of elemental phosphorus is embodied in chemicals used to manufacture other chemical products, mainly detergents.

Elemental chlorine and sodium hydroxide (NaOH, or caustic soda) are coproduced by electrolysis of sodium chloride (salt), mainly in the form of brine (Figure 7). In 1988, U.S. salt production was 35 MMT, of which 18.1 MMT was consumed by chlor-alkali producers. The United States produced 10.21 MMT of chlorine and 9.55 MMT of sodium hydroxide (Chemical and Engineering News, 1992).[28] The wastes from this process, mostly spent brines, amount to about 15 percent of the weight of the products, or 3 MMT. The mass balance is made up from about 5 MMT water on the input side and 0.3 MMT hydrogen gas released from the electrolytic cells. The latter is generally burned on site for energy recovery.

Major uses of chlorine in the United States in 1988 included chemicals manufacturing (76 percent), water and sewage treatment (5 percent), pulp and paper bleaching (14 percent), titanium dioxide manufacturing from rutile (3 percent), and miscellaneous, including silicon processing (2 percent). The use of elemental chlorine for bleaching in the pulp and paper industry has become a very contentious subject in recent years, due to the discovery of dioxin traces in bleached paper products. As a consequence, whether justified or not, this bleaching process has been largely phased out in Europe and may soon be phased out in North America. The likely alternative process is oxygen bleaching using chlorine oxide from sodium chlorate[29] or hydrogen peroxide.

Chemical end uses of chlorine in 1988 were as bleaches such as calcium and sodium hypochlorite (7.8 percent); other inorganics like phosgene (2 percent) and phosphorus trichloride (1 percent); ethylene dichloride, or EDC (40.5 percent); chlorinated methanes (9 percent); chlorinated ethanes (5 percent); epichlorohydrin (1 percent); ethyl chloride (3.5 percent); chlorinated benzenes (1.5 percent); chloroprene (1 percent); and miscellaneous (3 percent, including hydrochloric acid [HCl] used outside the industry). These total to about three-quarters of U.S. chlorine output.

One of the major chlorine intermediates is HCl. Nine percent of HCl is made by direct chlorination of hydrogen, but most HCl, about 91 percent, is recovered as a by-product of one of the chlorination processes, especially the chlorohydrin process for propylene oxide production. The latter process consumed 8.3 percent of U.S. chlorine output in 1988, but this chlorine was entirely recycled internally as HCl. The other major source of HCl is the process that converts EDC to vinyl chloride monomer (VCM). HCl is also consumed by a parallel process that produces EDC by hydrochlorination of ethylene. These two processes are deliberately combined. Other end uses of HCl include ethyl chloride (for tetraethyl lead production, now nearly phased out) and hypochlorite bleaches.

By far the biggest single intermediate is EDC, the intermediate leading to polyvinyl chloride (PVC). However, PVC accounts for only 24 percent of produced chlorine; part of the chlorine contained in EDC is reclaimed again as HCl. EDC also has other end products, such as trichloroethylene and perchloroethylene. Some is also exported. Another important intermediate is epichlorohydrin, an intermediate to epoxies; phosgene ($COCl_2$) is an intermediate to isocyanate pesticides and urethanes. Chlorinated benzenes are also intermediates to a variety of products.

Virtually all uses of chlorine are dissipative, with the major exception of PVC, which is used for structural purposes (e.g., water and sewer pipes, siding, window frames, calendered products, and bottles). PVC, from vinyl chloride monomer, accounts for 24 percent of U.S. elemental chlorine output.

Direct chlorination processes are relatively inefficient. Hence, recycling of waste streams is commonplace and a fairly large proportion of the input chlorine eventually becomes a production waste for manufacturing other downstream chemicals. For example, in the production of VCM from ethylene dichloride, the loss rate is around 3 percent (Manzone, 1993). About 60 percent of this waste stream consists of nonvolatiles, or heavy ends, that are recycled into chlorinated solvents; 40 percent is volatiles that are destroyed by incineration, with some recovery of hydrogen chloride. In general, a 3 percent loss rate for direct chlorination processes seems realistic.

The other halogens are bromine, fluorine, and iodine. The last of these is not used in significant quantities. Bromine consumption in the United States in 1988 was 0.163 MMT. Exact figures are unavailable, but the Bureau of Mines estimated that about 25 percent was used in drilling fluids, mostly as calcium and zinc bromides, and 15 percent was used in water treatment as a biocide for slime control. Fire retardants (tetrabromobisphenol-A and decabromodiphenyl oxide, known as halons) accounted for about 30 percent of total usage. About 15 percent was used in agriculture as a soil fumigant, methyl bromide, whereas 18 percent was consumed as ethylene dibromide, which is added to leaded gasoline as a scavenger to prevent lead deposition on valves. (This use was being phased out in 1988 and is now [1998] negligible.) An unknown but small amount was used to manufacture a red pigment (pigment 168) for metallized paint for automobiles. Evidently, around 60 percent was used in organic chemicals, half olefin based and half aromatic (e.g., phenol) based. All uses of bromine are dissipative.

Fluorine consumption in the United States was 0.551 MMT in 1988, but this included nonchemical uses such as fluxes for the steel industry. Synthetic cryolite for the aluminum industry may have used some hydrofluoric acid (HF), although for this purpose the U.S. industry is converting to the use of by-product fluosilicic acid from phosphate rock processing.

Most chemical uses of fluorine begin with HF, produced by reacting sulfuric acid and fluorspar (CaF_2). U.S. production of the acid in 1988 was 0.195 MMT, consuming roughly 0.24 MMT (S content) of sulfuric acid and generating calcium sulfate as a waste. Some HF was used in petroleum refining (as an alkylation catalyst), and some was used in uranium refining (uranium hexafluoride, UF_6). An increasing amount is being used in semiconductor manufacturing. However, probably most of the fluorine was consumed in the manufacture of chlorofluorocarbons (CFCs) and hydrochlorofluorocarbons (HCFCs), notably CCl_3F, or CFC-11 (0.1 MMT containing 10 percent F); CCl_2F_2, or CFC-12 (0.175 MMT containing 24 percent F); and $CHClF_2$, or HCFC-22 (0.148 MMT containing 34 percent F). Altogether, these three CFCs accounted for about 0.092 MMT flourine, or over 0.10 MMT HF. Other CFCs and the fluorocarbon polymers (e.g., Teflon©) presumably accounted for most of the rest.

In 1988, as noted above, 9.55 MMT of sodium hydroxide (caustic soda) were produced as a coproduct of chlorine production. We have no precise breakdown for 1988. Major uses of caustic soda in 1973 were in the chemical (46 percent) and pulp and paper industries (16 percent) for the preparation of alumina by the Bayer process and to make synthetic cryolite for aluminum manufacturing (6 percent); in petroleum refining (6 percent), dyeing of textiles (4 percent), and rayon manufacturing (3 percent); in the production of soap and detergents (3 percent) and cellulose acetate (2 percent); and for miscellaneous purposes (14 percent, including exports) (Lowenheim and Moran, 1975, p. 742). By 1988, the pulp and paper industry had increased its share to 24 percent (2.7 MMT), and the aluminum industry had consumed just under 4 percent (0.37 MMT), a slight decline.

Taking into account alumina and soap and detergent manufacturing, around 48 percent (4.5 MMT) of caustic soda was probably absorbed by the chemical manufacturing sector. Most sodium-containing inorganic chemicals such as sodium silicate or sodium dichromate start from less-expensive sodium carbonate (soda ash), rather than sodium hydroxide. Sodium is not a significant component of synthetic organic chemicals, with one exception: sodium salts of coconut oil acids (0.275 MMT) and tallow acids (0.46 MMT), which are major components of liquid detergents. (We cannot readily estimate the average sodium content of these, but we conjecture that it is less than 10 percent, which implies a 0.2–0.3 MMT Na_2CO_3 equivalent.) This is consistent with the 3 percent share noted above. Otherwise, virtually all of the produced sodium hydroxide was dissipated within the chemical sector itself, mainly for acid neutralization. Therefore, we estimate that dissipative losses of caustic soda within the chemical industry was about 4.2 MMT.

Soda ash (sodium carbonate) is another alkali sodium chemical that was once manufactured synthetically by the Solvay process. However, sodium carbonate is now extracted from brines and evaporite deposits from the western United States. Of U.S. production in 1988, 2.117 MMT were exported (net). Domestic uses of sodium carbonate amounted to 7.6 MMT, of which 1 MMT was taken from stocks. Traditionally, about half of sodium carbonate use is in glass manufacturing. Other uses include in alkaline cleaners (12 percent), pulp and paper (2 percent), flue-gas scrubbing (2 percent), and water treatment (1 percent). The glass industry uses about 0.28 MT of soda ash per metric ton of glass produced.

Around 22 percent (1.9 MMT) of soda ash is used in other mostly inorganic chemical manufacturing. Important sodium chemicals that use sodium carbonate as a feedstock include STPP, of which 1988 production was 0.497 MMT (0.124 MMT Na); sodium silicate (0.736 MMT); and sodium cyanide (0.154 MMT). All of these sodium-containing chemicals are dissipated in final use.

Aluminum chemicals, notably alumina and aluminum sulfate, were mentioned above. However, alumina, although classed as an inorganic chemical, is really a feedstock for aluminum refining and need not be discussed further here. Aluminum sulfate, produced by reacting alumina with sulfuric acid, is used in the paper industry.

Silicon chemicals of importance include sodium silicate, silicones, and silanes, which are used in the production of ultrapure polycrystalline or amorphous silicon for semiconductors. The major inorganic silicon chemical is sodium silicate. Organic silicon chemistry is highly specialized and there are few data.

However, it is estimated that production of silicone resins for synthetic rubber amounted to 0.0947 MMT in 1990. Figures for 1988 were probably similar.

Compounds of iron, chromium, copper, lead, titanium, and zinc also have important chemical uses, especially for pigments. Titanium dioxide has already been mentioned in connection with chlorine. It is the most important metallic pigment, used for most exterior paints as well as in paper. U.S. production in 1988 was 0.926 MMT, mostly from ilmenite. For each ton of TiO_2 produced, 1.2 tons of waste are generated, implying 1.11 MMT of waste from this source in 1988.

Iron oxide is a red pigment that comes largely from natural sources. Ferrous chloride and ferrous sulfate are by-products of steel pickling with hydrochloric and sulfuric acids, respectively. Ferrous chloride is used to some extent as a soil conditioner. The supply of ferrous sulfate, in low-grade forms, is much greater (2–4 MMT) than the demand for it. The compound is used to some extent to make iron oxide, to manufacture ferrites, as a catalyst, and in sewage treatment. However, much of it must be disposed of as waste.

Copper sulfate (0.0342 MMT) is the basis of most copper chemicals (fungicides, algicides, pesticides, catalysts, flotation reagents, etc.). It is made by reacting scrap copper with sulfuric acid. Its uses are dissipative, mainly in agriculture and wood preservatives.

Chrome-containing chemicals are mostly derived from sodium dichromate (0.145 MMT Cr_2O_3). Their largest use in 1988 was in wood preservatives, in combination with arsenic and copper. Use in preservatives accounted for 0.047 MMT of sodium dichromate equivalent, or 43 percent of 1991 demand (Roskill Ltd., 1991). Chromic acid is used for electroplating and metal treatment (15 percent of 1991 demand) and as the base for producing other chromium chemicals. Leather tanning accounted for 9 percent of U.S. dichromate demand in 1991. Tanning employs chromium (III) sulfate to protect the leather from attack by microorganisms. Chromium pigments (green, yellow) were a major use in the early 1980s, but dropped to just 8 percent of domestic demand by 1991. (Use as pigments in 1988 was about 0.014 MMT.) Virtually no copper or chromium was used in the manufacture of synthetic organic chemicals, except as catalysts. All chemical uses of chromium except for electroplating are essentially dissipative.

Lead sulfates and oxides are primarily pigments but are also the basis for other lead chemicals such as tetraethyl lead. They are still produced in fairly large, though decreasing, quantities. Lead oxides for pigments, including litharge, red lead, and white lead, accounted for 0.551 MMT, gross weight, of which 0.522 MMT was lead. Tetraethyl (TEL) and tetramethyl lead were once produced in very large quantities as gasoline additives. Their production and use have declined sharply since 1970 as a result of environmental regulation. Based on the ratio of ethyl dibromide to TEL in earlier years, we estimate that U.S. production of these additives in 1988 was approximately 0.035 MMT, almost all of which was lead itself. These additives are classed as synthetic organic chemicals. Fuel uses of TEL are obviously dissipative.

The following zinc chemicals were produced in 1988: zinc oxide (directly from ore by the so-called French process), 0.0345 MMT; zinc sulfate, 0.013 MMT; and zinc chloride, 0.085 MMT. Zinc oxide is used mainly in tire manufacturing and as a pigment. Zinc chloride is mainly used as an electrolyte in dry cells. Minor quantities of zinc were used in pesticides and to manufacture catalysts. Except for these uses, zinc is not consumed in the production of synthetic organic chemicals. All final uses are dissipative.

Synthetic Organic Chemicals[30]

Most organic industrial chemicals are based on petrochemical (hydrocarbon) feedstocks. There are three basic categories: (1) paraffins (alkanes), which are saturated aliphatic (straight or branched-chain) hydrocarbons, the most important of which are methane, ethane, propane, isobutane, and n-butane; (2) olefins (alkenes), unsaturated aliphatics with one or more double bonds (e.g., ethylene, propylene, butylene, butadiene); (3) cyclics and aromatics (e.g., benzene, toluene, xylenes, cyclopentane, cyclohexane, and naphthalene). There is a fourth, miscellaneous, group of nonhydrocarbons, including oxygenated compounds of organic origin, cellulose, fatty acids, and related chemicals.

Some of the primary feedstocks of alkanes and aromatics, totaling 32.44 MMT in 1988, were derived from natural gas liquids (22.46 MMT), refinery off-gas (1.12 MMT), and naphtha (8.864 MMT) (International Energy Agency, 1991). Separate data are no longer collected for ethane, propane, and butane, probably due to the prevalence of mixed streams generated and converted within the petroleum refining sector.

For our purposes, it is convenient to exclude the C_2–C_4 alkanes (ethane, propane, isobutane, and n-butane) from consideration, because virtually all organic chemical products, except methanol, are derived from the corresponding olefins. The latter, in turn, are almost entirely used for chemical conversion. Primary products for chemical conversion, including methane, C_2–C_4 olefins, C_5 and "other" aliphatics (including methane), and aromatics, consumed in the United States amounted to approximately 63.6 MMT in 1988, including net imports of 7.64 MMT, according to the Bureau of Mines (1991). The remainder was derived from petroleum refineries (32.44 MMT, see previous section) and natural gas (23.5 MMT, estimated).

Not all of this material was converted into petrochemical products. For example, U.S. refineries produced 10.61 MMT of benzene, toluene, and xylenes (known as BTX), but *Chemical and Engineering News* (1997) lists only 4.08 MMT as "chemicals"; the implication is that 6.5 MMT were used by the refineries as gasoline additives. In addition, refineries produce significant quantities of hydrogen from the dehydrogenation process. We have no exact figures, but chemistry suggests that hydrogen would have accounted for about 10 percent of the input mass, or 6 MMT. Most of this was presumably used for other refinery operations, especially hydrogen reforming and hydrogen desulfurization. Some of the input material, mostly gas, was burned to provide energy for the dehydrogenation and cracking furnaces. Again, we have no precise data, but 5 percent (3 MMT) seems a reasonable estimate.

A further 1–2 percent of feedstocks (0.6–1.2 MMT) may have been lost as VOC emissions. Even so, we cannot fully account for the outputs. For example, it is not clear from the data whether natural gas consumed for ammonia production is or is not included. (Urea, made from ammonia, is included as an organic chemical product.) Methanol is certainly one of the primary products, but most methanol is imported, and the domestic product (1.85 MMT in 1988) could not account for very much of the 8.8 MMT of "missing" feedstocks.

To simplify somewhat, we consider the "true" feedstocks to the organic chemical industry to be C_2–C_5 olefins (33.35 MMT, approximately), aromatics (BTX and naphthalenes, 5.78 MMT, excluding the BTX diverted to gasoline additives), and methanol (4 MMT). The grand total of hydrocarbon feedstocks and methanol appears to have been 43.1 MMT in 1988. Some of this production, especially C_4–C_5 and higher-order aliphatics, was not actually used to manufacture other chemicals. Some was used as an octane booster in gasoline or as a solvent; some was converted to hydrogen, mostly used in the refining process, or to carbon black.[31] However, most downstream synthetic organic chemicals are derived from the above-mentioned sources or the previously discussed inorganic intermediates. Olefins, in particular, are almost immediately and completely converted to polymers or other chemical intermediates such as alcohols and/or resins. In addition, we must account for miscellaneous organics such as glycerol (about 0.25 MMT, derived from animal or vegetable oils), fatty acids from vegetable oils used in liquid-detergent manufacture (0.735 MMT), and soluble cellulose used for cellulose acetate but not rayon (0.5 MMT).[32]

We also include specified fractions of the major inorganics discussed above. The latter include nitrogen chemicals, chlorine chemicals, sulfuric acid, and sodium hydroxide not used for other purposes and accounted for elsewhere. The last two reagents, in particular, are used in great quantities, but very little of the reactive element in either case is embodied in final products.

As regards chemicals, the situation is confused by imports, exports, and by-products at various stages. In the case of ammonia, over 90 percent goes to fertilizers, explosives, synthetic fibers, plastics, and other identified inorganics. We can account for 0.737 MMT (N content) embodied in organic chemicals, plus about 0.16 MMT of associated process losses in 1988. So, a total of 0.9 MMT N was consumed in making synthetic organic chemicals in 1988, mainly plastics and fibers. Total weight of ammonia used would therefore have been about 1.1 MMT. For sulfur, the amount embodied in organic products was very small (0.08 MMT), but the amount used dissipatively in the industry was at least 0.6 MMT and, thanks to the large amount unaccounted for, could have been as much as 1.9 MMT, of which as much as 1.45 MMT could have been in the form of sulfuric acid (Bureau of Mines, 1989). This would correspond to about 4.5 MMT of the acid. In the case of chlorine (see above), we saw that about 6.6 MMT were used within the chemical industry to make organic chemicals or final products. As regards caustic soda, it appears that 4.2 MMT were used for acid neutralization within the chemical industry. In addition, small amounts of fluorine (0.18 MMT), bromine (0.1 MMT), and phosphorus pentoxide (0.075 MMT) were used.[33]

Uncertainties in sulfuric-acid and caustic-soda requirements can be reduced, based on the knowledge that all acids and alkalis must be neutralized. Moreover, apart from small amounts of hydrochloric, nitric, and hydrofluoric acid, sulfuric acid is the dominant source of acidity (H+), while caustic soda and ammonium hydroxide are essentially the only sources of balancing hydroxyls (OH⁻). However, ammonium sulfate is virtually all recovered in fertilizer, and both N and S have already been accounted for. Thus, virtually all other sodium and sulfur inputs must end up in waste streams (largely as sodium sulfate), because they are not embodied in products. The basic neutralization reaction is

$$2NaOH + H_2SO_4 \Rightarrow Na_2SO_4 + 2H_2O.$$

The molecular weight of sulfuric acid is 98 g mole whereas the molecular weight of caustic soda is 40 g mole. Because the reaction requires two moles of caustic soda per mole of sulfuric acid, it would consume 1.22 mass units of sulfuric acid per unit of caustic soda. Thus, if 4.2 MMT of caustic soda was neutralized in the organic sector, we would have needed 5.15 MMT of sulfuric acid.[34]

Based on the above analysis, we argue that, in the synthetic organic chemicals sector, the ratio of sulfuric acid to caustic soda, in mass units, must be close to 1.22. Because the maximum amount of sulfuric acid available for organic synthesis processes, including acid unaccounted for, was 4.5 MMT, we conclude that not much more than 3.7 MMT of caustic soda could have been used in the same processes. (The remainder of the caustic, about 0.5 MMT, must have been used in the inorganic chemical sector.)

Adding these (Figure 8), we arrive at a grand total of 61 MMT of produced chemical inputs to organic synthesis in 1988. Oxygen is needed for a number of downstream oxidation processes, such as production of ethylene and propylene oxides; ethanol, isopropanol, and butanol; phthalic anhydride; terephthalic acid; and oxy-chlorination of ethylene to EDC.

A survey of the major products of the synthetic organic sector reveals that the oxygen content of final product chemicals averages close to 10 percent, which would amount to a total of about 4 MMT oxygen (O). We can account for 1.85 MMT O embodied in the input methanol. There is also some oxygen in cellulose and fatty acids. Some oxygen is carried into the reactions by the oxidizing agents nitric and sulfuric acid (HNO_3 and H_2SO_4). But nitric acid is itself produced by

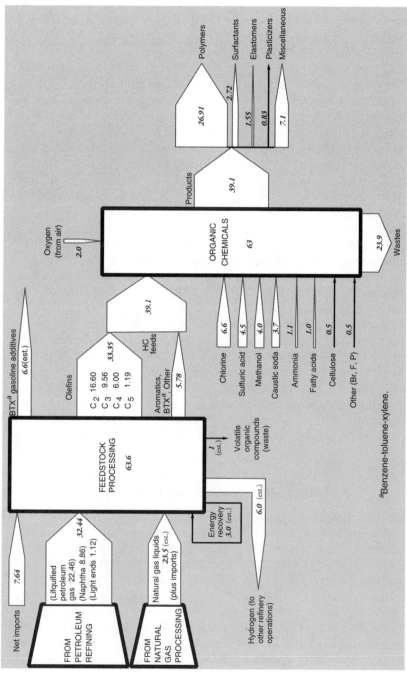

FIGURE 8   Mass flows in the U.S. organic chemical industry, 1988 (million metric tons). Calculated by the authors from various sources.
[a]Benzene-toluene-xylene.

oxidation of ammonia, and sulfuric acid is the oxidized form of sulfur (dissolved in water). We estimated above that 1.45 MMT S, or 4.5 MMT of sulfuric acid, were utilized in organic synthesis in 1988. This would include 2.9 MMT O. We must also allow for 3.7 MMT NaOH used in organic synthesis processes, of which 1.4 MMT was oxygen. Altogether, this adds up to 6.15 MMT O, which is somewhat more than the amount actually embodied in final products.

In principle, no additional oxygen is needed, except to combine with carbon and/or hydrogen in the feedstock to generate energy to drive the endothermic syntheses. In practice, however, very little of the oxygen embodied in caustic soda or sulfuric acid ends up in the products. Rather, it ends up as water in the neutralization reaction. Thus, to arrive at a final oxygen content of 3.9 MMT, as indicated, we must assume that at least 2 MMT have been extracted from the atmosphere in various processes. This does not include the oxygen needed to oxidize all the missing mass to its final form, however.

Adding these, we arrive at a grand total of 61 MMT of produced chemicals and 2 MMT of oxygen as inputs to organic synthesis in the United States in 1988, not including oxygen that is used for combustion purposes and finishes as carbon dioxide. The major outputs, in terms of sales, of the organic chemicals industry amounted to 39.1 MMT in 1988 (and 39.5 MMT in 1989), not including urea. The categories are listed in Table 6.

Subtracting the weight of identified products from the weight of inputs, we find that the missing mass was around 23.9 MMT, not including the mass of any oxygen combined with carbon as process wastes. The situation is summarized graphically in Figure 8. The approximate composition of the waste stream can be estimated from the composition of the inputs; however, it is clear that water, sodium sulfate, and carbon dioxide must account for most of the waste. The remainder consists of various other salts, including some NaCl, and VOCs. A more detailed breakdown would be possible if we knew the inputs more accurately.

Primary Metals Smelting and Refining[35]

We have conceptually divided the processes of mining, concentration (winning), reduction (smelting), and refining. There are four stages of separation or recombination. The first two, being physical in nature, are assigned to the mining sector or the quarrying sector. They were discussed above. The last two, being chemical in nature, are assigned to the primary metals sector. At each separation stage, wastes are left behind and a purified product is sent along to the next stage. In principle, the wastes can be determined by subtracting outputs from inputs.

Unfortunately, from the analytic point of view, appropriate published data are rarely available. There are significant imports and exports of concentrates and crude metals (and even some crude ores), but trade data are often given in terms of metal content rather than gross weight. Domestic data are also incom-

TABLE 6   End-Use Organic Chemical Products: U.S. Production and
Sales, 1989 and 1990 (million metric tons)

	Production		Sales	
Chemical Product	1989	1990	1989	1990
TOTAL		503.3880		39.1078
Dyes	0.174	0.1170	0.146	0.1040
Organic pigments	0.050	0.0530	0.043	0.0450
Medicinals	0.130	0.1440	0.204	0.1070
Flavor and perfume materials	0.064	0.0600	0.038	0.0370
Rubber processing chemicals	0.176	0.1790	0.129	0.1360
Pesticides	0.572	0.5570	0.461	0.4420
Thermosetting resins		4.3095		3.1770
Thermoplastic resins		25.2013		22.0939
Polymers for fibers		2.3585		1.3791
Polymers (water soluble)		0.3097		0.2647
Elastomers		2.2331		1.5551
Plasticizers		0.8907		0.8265
Surfactants		5.8487		2.7181
Antifreeze	0.920	0.9009		0.9000
Chlorofluorocarbons	0.417	0.3083		0.3000
Solvents				1.2000
Chelating agents		0.1372		0.1015
Fuel additives		4.2247		1.9356
Lube oil and grease additives		0.3872		0.3436
Textile chemicals (excluding surfactants)		0.0224		0.0198
Miscellaneous chemicals		2.0966		1.4219

SOURCE: International Trade Commission (1992).

plete, due to information withheld for proprietary reasons. Thus, in a number of cases, we have been forced to work back from smelting or concentration process data to estimate the input quantities of concentrates. Our summary was given in Table 3. It excluded ferroalloys, of which U.S. production was about 1 MMT, because of the extreme complexity of the subsector.

Inputs to the U.S. primary metals sector consist of concentrates (either produced in the mining sector or imported), fuels, fluxes, and processing chemicals. Because we have accounted for inputs to and wastes from fossil fuel combustion in an above section, those quantities are not included in our accounting of wastes from primary metals smelting and refining. CO is a major pollutant of smelting processes, but it results from partial oxidation, which is later completed in the atmosphere. (Thus, the materials-balance approach is not applicable for estimating CO emissions.) Major purchased inputs, other than concentrates, are fluxes. The most important are limestone and dolomite. In 1988, approximately 9.6

MMT of limestone and dolomite were used, mostly in blast furnaces. Data published by the Bureau of Mines indicate that these materials were calcined on site and consumed as lime (4.8 MMT). Other inputs reported by the Bureau of Mines include salt (0.33 MMT), manganese ore (0.123 MMT), and fluorspar (0.137 MMT).

The production of primary metals from concentrates is normally accomplished by carbothermic reduction (smelting with coke) or electrolysis. By far the major product by weight is pig iron. U.S. blast-furnace output in 1988 was 50.9 MMT. This material has an iron content of 94 percent and it is almost entirely used for carbon steel production by the basic oxygen process. (There was a small amount of open-hearth production in 1988, which has now ended. Electric steel "minimills" use scrap almost exclusively.)

Blast furnace inputs in 1988 included about 3 MMT of scrap iron and steel, whereas sinter also utilized about 6 MMT of upstream reverts (dust, mill scale). Therefore, accounting for the virgin ore is somewhat complex. However, the iron content of U.S. ores in 1988 was reported as 57.515 MMT. Blast furnace inputs (pellets) averaged about 63 percent iron, 5 percent silica, 2 percent moisture, and 0.35 percent other minerals (phosphorus, manganese, alumina). The remainder was oxygen.

In the reduction process, the oxygen combines with carbon (actually carbon monoxide) from the coke. About 0.5 MT of coke was used per metric ton of pig iron, along with 0.142 MT of miscellaneous materials, mostly fluxes (lime and limestone) for the sinter plants and to make the molten slag flow easily. Slag consists of the silica and other nonferrous minerals in the sinter and pellets and the materials in the fluxes. Total iron blast furnace slag production in the United States was 14.2 MMT, or 0.28 MT of slag per metric ton of pig iron. However, slag is no longer considered a waste, because virtually all slag produced is marketed for a variety of uses. Subsequent refining of pig iron and scrap iron to carbon steel is done in a later refining stage, normally the basic oxygen furnace. Steel furnaces produced an additional 5 MMT of slag in 1988.

As noted, the oxygen in the iron-bearing concentrates reacts in the blast furnace with carbon monoxide. The reduction process requires excess CO, so the emissions (blast-furnace gas) consist mostly of unreacted CO. Although combustible, it is of relatively low heating value. Currently, most blast-furnace gas is used elsewhere in the integrated steel complex as fuel (e.g., for preheating blast air), although some is used as fuel by nearby electric power plants. The capture of gaseous emissions from blast furnaces is not 100 percent efficient, so some CO escapes. However, considering the iron/steel process as a whole, all of the carbon (from coke) is eventually oxidized to CO_2. In 1988, the steel industry accounted for 182 MMT of CO_2 from coke, which is included in the grand total from fossil fuel combustion, discussed below. (In addition, the steel industry used some other hydrocarbon fuels.)

Coke ovens and steel-rolling mills are significant sources of hazardous wastes, even though the coke-oven gas is efficiently captured for use as fuel, and about 0.055 MMT of ammonium sulfate (N content) is produced as a by-product. This material is used as fertilizer. Coke is cooled by rapid quenching with water, and some tars, cyanides, and other contaminants are unavoidably produced. Unfortunately, materials balances cannot be used to estimate these wastes. However, they probably constitute a significant fraction of both water and airborne wastes from the primary iron and steel sector.

Also, in the rolling process, steel is cleaned by an acid bath (pickling), resulting in a flow of dilute waste water containing ferrous sulfate or ferrous chloride, depending on the acid used. The excess acid is usually neutralized by the addition of lime. In 1988, about 0.215 MMT of 100 percent sulfuric acid (0.074 MMT S content) was used for this purpose, producing 0.25–0.30 MMT of ferrous sulfate mixed with calcium sulfate. Ferrous sulfate can, in principle, be recovered for sale to the water treatment industry. However, the market is insufficient to absorb the quantity potentially available, and most of it ends up as waste.

Light metals, mainly aluminum and phosphorus, are reduced electrolytically. The oxygen in the alumina reacts with a carbon anode made from petroleum coke. The reaction emits 0.65 MT of CO_2 per metric ton of primary aluminum produced. In addition, primary aluminum plants emit about 0.02 MT of fluorine per metric ton of aluminum, partly as HF and partly as particulates, the latter due to the breakdown of cryolite (the electrolyte used in the process, an aluminum–sodium fluoride) at the anode. Total airborne emissions (3,944 MMT) from primary aluminum production in the United States were, thus, 2,564 MMT of CO_2 (already counted), 0.08 MMT of fluorides, and about 0.17 MMT of particulates (Al_2O_3).

In the case of heavy metals from sulfide ores (copper, lead, zinc, nickel, molybdenum, etc.), the smelting process is preceded by, but integrated with, a roasting process whereby the sulfur is oxidized to SO_2. Roughly 1 MT of sulfur is associated with each MT of copper smelted, 0.43 MT of sulfur per MT of zinc, and 0.15 MT of sulfur per MT of lead. Ninety percent of this sulfur is captured and immediately converted to sulfuric acid. In 1988, 1.125 MMT (S content) of by-product sulfuric acid were produced at U.S. nonferrous metal refineries, as follows: copper (0.946 MMT S), zinc (0.136 MMT S), and lead/molybdenum (0.043 MMT S). In terms of sulfuric acid (100 percent H_2SO_4), 3.54 MMT of by-product acid were produced.

In the case of copper, most of the acid (1.2 MMT) was used by mines for heap leaching copper. Leaching now accounts for about 30 percent of copper concentrates produced in the United States. Leached copper sulfate is subsequently reduced electrolytically, without an intermediate smelting stage. In the case of copper smelting, typical concentrates fed to the roaster/smelter consist of about 35 percent Cu, 35 percent S, and 30 percent other minerals. In addition, about 0.25 MT of limestone flux is added per ton of blister copper. Thus, slag production amounts to roughly 0.55 MT per metric ton of primary copper, or 0.77 MMT in 1988.

In the case of zinc, a typical concentrate would be about 55 percent Zn, 27 percent S, and 16 percent other minerals. For lead, the corresponding numbers appear to be about 60 percent Pb, 9 percent S, and 21 percent other. Thus, assuming flux per unit of slag ratio to be the same as for copper (1.2:1), slag output should have been roughly 0.3 MT per metric ton for zinc and 0.38 MT per metric ton for lead. This implies total slag output of 0.06 MMT for zinc smelting and 0.14 MMT for lead smelting. Total slag production for the three main nonferrous metals was thus roughly 1 MMT. Carbon monoxide and carbon dioxide emissions are not known exactly, but they are quite small in comparison with other sources. The waste numbers for other metals are relatively insignificant.

Altogether, based on mass-balance considerations, we estimate smelting and refining wastes for primary metals, including CO_2, to have been 43.4 MMT in 1988, including the weight of limestone, manganese, calcium fluorite, and other materials used in the blast furnaces and refineries. (This includes about 14.2 MMT of iron, or blast-furnace, slag, although most of this material is marketed commercially, mainly for road ballast.) In addition, there were about 5.2 MMT of steel-furnace slag, a denser material with a fairly high iron content.

As noted above, much of the sulfur in sulfide copper, lead, zinc, and molybdenum ores is also recovered for use and sold as sulfuric acid (1.125 MMT S content in 1988). Subtracting the blast furnace slag and the by-product sulfuric acid, we get 28.1 MMT residual waste. Of this, only about 1 MMT was solid nonferrous slag; the rest was the oxygen content of the original ores combined with carbon (from coke), released as CO_2.[36] We have not included the wastes from coking, which we have not estimated. The major airborne emissions other than CO_2 are probably CO and particulates. In both cases, blast furnaces are the major sources. The coking quench waters and some spent acids used for pickling constitute the major waterborne wastes.

Mass flows and wastes for the metallic mineral processing industries and the metallurgical industries, taken together, are summarized in Figure 9 (ferrous) and Figure 10 (nonferrous). These values are normalized to U.S. production of the refined metals. Some of the flows are imputed from others. For example, pig iron (94 percent Fe) contains roughly 6 percent C, which implies a carbon content of 3 MMT. The oxygen required to burn this carbon away was therefore approximately 8 MMT, implying a CO_2 output of 11 MMT for steel production in 1988. In the case of iron blast furnaces, we assumed that all of the input coke, less the 3 MMT of C embodied in pig iron, was converted to CO_2. This consumed 63.4 MMT O_2 and generated 87.1 MMT CO_2. However, some oxygen was captured from the iron oxide in the ore. So, balancing inputs and outputs, we calculate that the additional oxygen taken from the air must have been 31.5 MMT. Scrap flows are very approximate, partly because scrap industry statistics are poor and partly because we have lumped stock adjustments and scrap flows for convenience.

For comparison, Science Applications International Corporation (1985) estimated the 1983 nongaseous wastes from iron and steel production at 6 MMT and

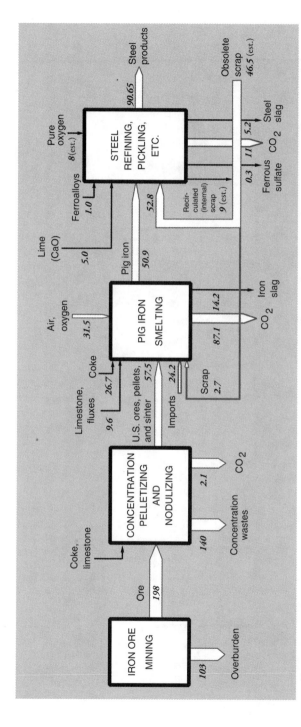

FIGURE 9   Mass flows in the U.S. iron and steel sectors, 1988 (million metric tons).   Calculated by the authors from various, sources, including Bureau of Mines (1988, 1989) and United States Environmental Protection Agency (1991). NOTE: Scrap consumption in iron and steel production is probably underestimated by up to 4 million tons. Recirculated scrap may be underestimated by a similar amount.

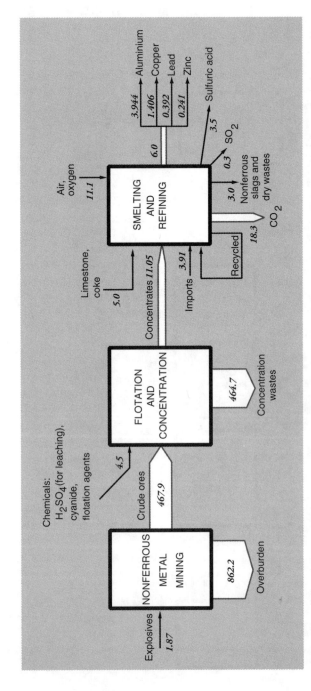

FIGURE 10 Mass flows in the U.S. nonferrous metals sector, 1988 (million metric tons). Calculated by the authors from various sources, including Bureau of Mines (1988, 1989).[*]

from nonferrous metals at 6.5 MMT. Their estimates were not specifically designated as dry, so some water content can be presumed. The EPA estimates airborne emissions from the primary metals sector as a whole to be 2.8 MMT, including particulates and CO but not including CO_2 (United States Environmental Protection Agency, 1991). Both sets of estimates are roughly consistent with our calculations.

Stone, Clay, and Glass[37]

As discussed above, overburden and concentration losses for nonmetallic minerals other than phosphorus in 1988 amounted to 47.1 MMT and 36.5 MMT, respectively. The outputs of the stone, clay, and glass sector include refractories, glass, and portland cement. All three are durable products used in structures or long-lived products. We do not have detailed quantitative data on glass production. As for clays, domestic production in 1988 was 43.9 MMT. Production of clay and refractory products, including clay used in Portland cement, was virtually identical. Some of these uses (e.g., fillers, binders, absorbents, drilling mud, filters) are essentially dissipative.

Portland cement manufacture is an important industry. Total U.S. production in 1988 was 66.5 MMT. Most of the input materials were natural minerals already discussed (limestone being by far the most important, 73 MMT), but small quantities of fly ash and blast-furnace slag also were used. Total nonfuel inputs were 111 MMT, and the mass lost in cement manufacturing was 44.5 MT. The major fuel consumed was coal (9.5 MMT), although some plants used oil or gas. Emissions are primarily CO_2 and particulates. The total weight of emissions from fuel was already counted above. However, CO_2 emissions from limestone calcination created an additional 35 MMT, or about 10 MMT C content, in waste. This still leaves nearly 10 MMT of missing mass. It is likely that some of this consists of particulate emissions from cement kilns, although this is probably not the complete explanation. There may be some internal recycling.

Lime (CaO) is made by calcining limestone. In 1988, U.S. production was 13.2 MMT (consuming 29.5 MMT of limestone and releasing 16.3 MMT CO_2 to the atmosphere). Uses of lime are extremely diverse and not well documented. The use of lime to treat stack gas was mentioned above. In many cases, limestone can substitute for lime (e.g., in glass manufacturing, soil stabilization, desulfurization). It must be emphasized that calcination of limestone releases CO_2 at a rate of 1.2 MT per metric ton of CaO.

It appears that the major waste emissions from this sector, exclusive of losses in quarrying and concentration, are primarily related to combustion of fossil fuels and calcination of limestone and gypsum, which yields CO_2. However, Science Applications International Corporation (1985) estimated dry wastes from the sector to be more than 18 MMT in 1983. EPA's latest estimate of airborne emissions, primarily particulates and CO, is about 1 MMT (United States Environmental Protection Agency, 1991).

We have no estimate of water use by the stone, clay, and glass sector. However, EPA estimated total wet wastes from the sector to be 560 MMT (United States Environmental Protection Agency, 1986). This seems quite high, given that most processes in the sector are dry.

Fossil Fuel Consumption[38]

Combustion of fossil fuels produces a variety of wastes. This is particularly true for coal. On the average, U.S. coal has a sulfur content of 1.9 percent; coal burned by electric utilities averages 2.3 percent sulfur, whereas coking coal is 1 percent sulfur. The latter is mostly recovered as ammonium sulfate. Coal burned in the United States emits about 16 MMT of sulfur (32.1 MMT SO_2). Most of this sulfur dioxide is released to the atmosphere.

In 1988, 1.24 MMT of lime (CaO) and 1.035 MMT of limestone ($CaCO_3$) were sold for use in removing sulfur from furnace stack gases. The limestone was equivalent to 0.495 MMT of lime. Because CaO has a molecular weight of 56 and SO_2 has a molecular weight of 64, the total amount of limestone and lime used in scrubbers accounted for only 1.96 MMT of sulfur dioxide, or about 6 percent of the total emitted. None of the sulfur from coal burning was recovered for further use. (It is disposed of in landfills as a mixture of wet calcium sulfite $CaSO_3$ and calcium sulfate $CaSO_4$.) EPA estimated that flue gas desulfurization by utilities produced 14.4 MMT of solid wastes in 1984 (United States Environmental Protection Agency, 1988, 1991). The mineral content of these wastes, even in 1988, was evidently no more than 3.7 MMT. The remainder was presumably water of hydration. (The mineral gypsum has the formula $CaSO_4 \cdot 2H_2O$.)

If all the sulfur in U.S. coal were to be captured by wet scrubbers using lime, total U.S. lime production would triple to 26 MMT, which would require an additional 55 MMT of limestone to be quarried. All of it would, of course, be converted almost directly into a waste stream.

Coal contains a small but significant percentage of fuel-bound nitrogen (about 1 unit per 68 units of carbon). Most of this is emitted as nitric oxide (NO) but some may be emitted as nitrous oxide (N_2O), one of the greenhouse gases. However, experts disagree about the amount of nitrous oxide produced by this process. More important, coal combustion in high-temperature boilers, used to generate electric power, produces a significant quantity of NO_x emissions, about 10 MMT/yr (United States Environmental Protection Agency, 1986). Virtually all anthropogenic NO_x (about 20 MMT/yr in 1980 and probably a similar amount in 1988) is attributable to the burning of fossil fuel.

Coal also contains significant quantities of mineral ash, equivalent to mineral shale. The average ash content of U.S. coal, as burned, is approximately 10 percent (Torrey, 1978). Actually, utilities alone seem to have collected and disposed of 62 MMT of ash in 1983. Assuming constant proportions of ash in coal used and complete ash recovery, the weight of disposed ash would have risen to 76 MMT by 1988, which would account for almost all of the ash in the utility coal. However, although the efficiency of recovery of fly ash from electrostatic precipitators is in the neighborhood of 99.8 percent for the most modern units, some utilities are not so well equipped. Fly ash not captured in 1988 probably amounted to at least 1 MMT. The ash content of coking coal, which is selected in part for its low ash content, ends up in metallurgical slag. The ash content of coal used as a fuel in the cement industry ends up as part of the cement itself. In fact, the cement industry also uses a small amount of fly ash as a raw material. Coal ash contains significant quantities of heavy metals. Although most fly ash is captured, the waste ash must be disposed of somehow. Moreover, the more volatile trace metals such as arsenic and mercury still escape as vapor and recondense downwind of the stack.[39]

Finally, the carbon in coal, along with the carbon in other fuels, is converted by combustion into CO_2. The Carbon Dioxide Information Analysis Center (1990) at Oak Ridge National Laboratory estimated that the carbon content of these fuels was 1,288.6 MMT, or 84.7 percent. Of this, 493 MMT was from solid fuels (coal), generating 1,810 MMT CO_2. This includes the CO_2 from carbothermic reduction processes using coke.

With the exception of electric power generation, most fuels are petroleum products or natural gas. Natural gas is mostly used for domestic purposes and space heating, although some is used in industry. Petroleum products are mainly used for transportation, although some heavy oils are used for industrial boilers or electric power production. The transportation system is of interest because there are so many complex mass flows involved, other than the straightforward consumption of fuel. We summarize this system, for private automobiles only, in Figure 11.

The sum total of all fossil fuels consumed in the United States in 1988 was 1,521 MMT (Table 5). We assume that all of the fossil fuel carbon was converted to CO_2 (4,726 MMT in 1988), not including calcination processes (lime and cement manufacturing), which are counted separately.

Combustion processes also result in some releases of methane to the atmosphere, but more methane escapes to the atmosphere during production and transmission. One study allocates 11.86 MMT of methane releases in the United States in 1988 to all of these activities together (Subak et al., 1992). The study does not provide a breakdown for the United States among production, transportation, and combustion. However, for the world as a whole, the breakdown was coal mining (62 percent), oil and gas extraction (14.8 percent), gas distribution (17 percent), firewood combustion (4 percent), and other fossil fuel combustion (2.3 percent). For the United States, firewood combustion would be a negligible source of methane, coal would be less important than it is globally, and natural gas would be more important than it is globally.

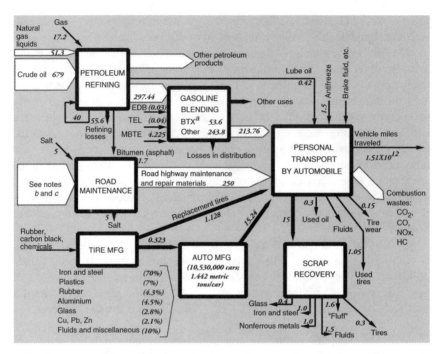

FIGURE 11   Mass flows associated with private automobiles in the United States, 1988 (million metric tons). Calculated by the authors from various sources, including International Energy Agency (1991), Bureau of Mines (1988), and Ecoplan International (1992).
[a]Benzene-toluene-xylente.

[b]Thrity-three percent of total road/highway expenditure is for maintenance and repair. Materials use per dollar is equal for new construction and for maintenance and repair. Fifty percent of all road/highway depreciation is attributable to automobile use.

[c]Total estimated materials used for road/highway construction, repair, and maintenance: bitumen, 16 MMT: Portland cement, 10 MMT; steel, 35 MMT; slag, 15 MMT; sand and gravel, 600 MMT; and crushed stone, 840 MMT.

SUMMARY

It may be interesting to summarize our results by waste category as well as by industry. Overburden moved by mining, mostly stripping, amounted to over 6,800 MMT in 1988. By contrast, topsoil loss in agriculture was on the order of 1,500 MMT. (In addition, the construction industry probably moves comparable amounts of topsoil.) Mineral concentration activities, mostly by froth flotation, produced waste (tailings) on the order of 900 MMT (600.6 MMT metals, 140.7 MMT nonmetals, 47 MMT coal cleaning, 57 MMT drilling wastes, 57.2 MMT phosphates and alumina) dry weight. In addition, about 3,600 MMT of water was used for flotation, most of which was evaporated in ponds, leaving semisolid sludges. Waste water discharged into rivers and streams by the mining industry amounted to 2,840 million gallons per day, or 3,900 MMT for the year.

By contrast, the weight of solid wastes from metallurgical conversion and fossil fuel combustion processes, including metallurgical reduction (smelting), amounted to only about 146.4 MMT. We have included in this category 76 MMT of fly ash and bottom ash from thermal power generators and 14.4 MMT of flue gas desulfurization waste but excluded 14.4 MMT of iron/steel slag that have commercial uses.

Organic pretreatment wastes are more difficult to account for. Crop residues, mostly recycled to land, amounted to about 360 MMT. Timber residues from logging operations amounted to 155 MMT, mostly burned or recycled to forest soils. (In some countries, both agricultural and timber residues are collected and burned as fuel, but this is relatively rare in the United States.) About 180 MMT (50 percent dry) of animal wastes—manure, urine, and dead animals— were produced, of which an estimated 155 MMT were probably recycled to cultivated land and the remainder lost in other unspecified ways. An additional 110 MMT of organic wastes (50 percent dry) were generated in the food processing sector and disposed of in various unspecified ways, including waterways. About 5 MMT were lost in the wood products and pulp and paper sectors.

Gaseous combustion products constituted another very large waste stream. We estimate gross emissions of 5,046 MMT CO_2, of which 4,726 MMT were from fossil fuel combustion. However, thanks to a takeup of 1,002 MMT by the agricultural sector and 368 MMT by the forestry sector, net emissions of CO_2 were 3,759 MMT. (By the same token, industrial activities, mainly fuel combustion, consumed 5,393 MMT of oxygen, whereas agriculture and forestry produced 891 MMT of oxygen, for a net consumption of 4,732 MMT.) Other gaseous emissions included 32 MMT SO_2 from coal and EPA's estimate of 20 MMT NO_X from fossil fuel combustion. Our methodology is not well suited to estimating fugitive or particulate emissions. However, we note that petroleum refineries may have emitted as much as 4.3 MMT of hydrocarbons that are not accounted for anywhere else.

Process water contaminated by acids or other wastes was also emitted in significant quantities by the petroleum refining and metallurgical sectors. The quantitative values discussed above are summarized in Table 7.

TABLE 7 Summary of Estimated U.S. Dry-Waste Streams, Excluding Water, 1988 (million metric tons)

Sector	Lost Mass				Process Wastes (ash, slag, etc.)	Emission to Air				
	Dry Organic		Dry Inorganic			H_2O	CO_2	O_2	CH_4, VOC	N,S
	Combustible	Noncombustible	Soil Overburden	Concentration Wastes						
Agriculture	+360[a] +155[b]	25	1,500			-1,027 +202	-1,002 +213	-156 +685	0.7	12 (N)
Food processing	?	?				66 (+19)[c]	6.8 (+29.4)[c]	(-21.3)[c]		
Forestry, pulp, paper	4.9				18.1	-545 +314	-368 +132	-74 +206		
Mining			965[d] 275.7[e] 5.600[f] 6–7[g]	600.6[d] 140.7[e] 47[f] 57[g]					9.8	
Petroleum refining[h]						(76.5)[c]	(165)[c]	(-120)[c]	2.0	

Chemicals		57.2					1.0	
Primary metals		22.9[i]						0.3 (S)
		0.3[j]						
		3.0[k]						
Stone, clay, glass		10			51.3			
Fossil fuel and electric power	520+?	92	2,088	4,726	-5,393		6	10 (N) 14 (S)
TOTAL	25+?	8,346.7	902.5	146.3	1,098	3,759.1	-4,732	19.5 36.3

[a] Crop residues, normally recycled to land.
[b] Animal manure, normally recycled to land. Not including emissions from free-ranging animals.
[c] Values in parentheses are emissions or consumption (−) from energy recovery. These values are not included in totals.
[d] Metal ore mining in the United States.
[e] Nonmetallic mineral mining in United States, excluding alumina and phosphate concentration wastes.
[f] Coal mining, not including coke-oven emissions.
[g] Petroleum and gas drilling.
[h] Including natural gas processing and transmission. H_2O, CO_2, and O_2 emissions (consumption) based on combustion of process wastes for energy.
[i] Process wastes of the organic chemical sector are partly combustible and partly incombustible. Detailed breakdown not available.
[j] Iron and steel slag are by-products with many uses. Concentration wastes are included with the mining sector.
[k] CO_2 emissions are included with fuel combustion

NOTES: Plus (+) signs refer to production. Minus (−) signs refer to consumption. VOC = volatile organic chemical.

NOTES

1. Where specific citations are not given, basic data for this section are from Hoffman (1991), United States Department of Agriculture (1992), and Bureau of Census (1988).

2. For the sake of clarity, it should be noted that "harvested output" of corn—by far the dominant grain—refers to shelled corn, not ears. The husks and ears are left behind on the farm, along with stalks. Similarly, wheat straw and chaff are separated from the wheat grains by the harvester and left behind.

3. To calculate the totals, it is necessary to sum up individual figures given by the United States Department of Agriculture (USDA) in a variety of different units, some volumetric and some in mass terms. Unfortunately, although the USDA does provide conversion coefficients, it does not calculate aggregated totals, except for grains.

4. It should be noted that although the totals of commercial high-protein feeds remain comparatively stable from year to year, the composition varies significantly. We are unable to account in detail for the exported "feeds and fodders" (11.4 MMT in 1988), which apparently originate in the processing sector (Bureau of the Census, 1991).

5. These estimates do not represent either the "fresh" weight of manure—which is relatively wet— or the "dry" weight of its solid content. Being based on inputs, the manure is assumed to have the same water content as the feeds (i.e., about 50 percent). In the case of cattle and pigs, actual fresh weight of animal manure is about four times greater than a similar volume of feed and at least in the case of other animals is at least twice as heavy.

6. Unless otherwise specified, production, export, and import data in this section are from Bureau of the Census (1988) tables 1148, 1149, 1156, 1163, 1166, 1167, 1168, 1173, 1175, and 1177. Data on per capita consumption of foods are given in tables 207 and 208. Beverages were not taken into account.

7. See United Nations Statistical Office (1988) tables on "Lard," ISIC 3111-31, and "Oils and fats of animals, unprocessed," ISIC 3115-07.

8. See United Nations Statistical Office (1988) table on "Hides, cattle and horses, undressed—total production," ISIC 3111-311. This refers to fresh weight, prior to tanning.

9. See United Nations Statistical Office (1988) tables on "Poultry, dressed, fresh (Total Production)," ISIC 3511-10, and "Poultry, dressed, fresh (Industrial Production)," ISIC 3511-101. For mysterious reasons, the latter figure is slightly larger.

10. Unless otherwise specified, data in this section are from Bureau of the Census (1991).

11. These data are essentially consistent with Ulrich (1990); however, Ulrich's table 51 appears inconsistent with Ulrich's table 7 as regards imports and exports of pulp. Table 7 includes pulp imputed to downstream paper and paperboard products. We account for imports and exports of downstream products separately.

12. Imports of "pulp products" shown in Ulrich (1990, table 4) apparently refer to pulp itself and to the pulp equivalent of paper and paper products, not pulpwood. The United States was a net exporter of pulp and a net importer of paper products.

13. Of this, 5.4 MMT were exported and 0.6 MMT was used for other purposes (Bureau of the Census, 1991, table 1194).

14. Actually, this is a lower limit, because it includes only inorganic materials (kaolin, alum, etc.) that we have been able to account for explicitly from published sources.

15. An attractive future possibility is to ferment or otherwise convert the hemicellulosics (sugars) to ethanol. Until now, all known fermenting agents produce an enzyme, lactate dehydrogenase, that breaks down the hemicellulosics into a mixture of ethanol and lactic acid. A new discovery at Imperial College, London, may change this situation. It is a mutant strain of the fermenting bacterium *Bacillus stearothermophilus*, which lacks the lactate enzyme and thus converts hemicellulosics directly to ethanol, without the usual mixture of lactic acid. Unfortunately, the mutation appears unstable and the organism reverts back to the original form, which produces

the enzyme. Therefore, the current challenge is to bioengineer a strain that completely lacks the gene.

16. This does not take into account the bark, which constitutes about 11.5 percent of the raw weight of roundwood (United States Congress, Office of Technology Assessment, 1984, table 5). We have completely omitted bark from our calculations by assuming that a cord of roughwood is equivalent to 80 cubic feet of debarked (peeled) roundwood. This suggests that about 10 MMT of bark would be produced by debarking operations, which precede pulping proper. Based on 48 percent moisture content, 10 MMT raw weight might be consistent with 5 MMT dry weight. (Bark is burned as "hog fuel.")

17. U.S. production of sodium chlorate in just the third quarter of 1992 was 0.131 MMT, which implies an annual rate of over 0.5 MMT, twice the level of 1988. In the same quarter, apparent consumption was 0.203 MMT, of which 37 percent was imported (United States Department of Commerce, 1992). The explanation is that chlorine dioxide has been very rapidly substituting for elemental chlorine as a bleach for pulp. Most of the increase in U.S. demand since 1988 is for conversion to chlorine dioxide.

18. U.S. production of paper products in 1988 consisted of 5.364 MMT of bleached newsprint (made from mechanical pulp), 19.59 MMT of printing and writing paper (coated and bleached), and 44.57 MMT of "other machine made paper and paperboard," of which 36.056 MMT consisted of Kraft paper and paperboard (Bureau of the Census, 1991).

19. Data on materials handled are from Bureau of Mines (1989, Volume 1, tables 10, 11). Other data in this section on metals and minerals come from individual chapters in the same source.

20. In the case of iron, concentrates for blast furnaces (pellets and sinter) are treated differently. Pellets are produced at the mine, whereas sinter is included in the smelting sector rather than in the mining sector. For consistency, we adopt this convention. In the case of aluminum, the concentration stage is taken to be the chemical conversion of bauxite ore into pure aluminum oxide, or alumina. This process is conventionally included in the inorganic chemical industry. Phosphate rock concentration, yielding fertilizer-grade superphosphate, is included in the fertilizer industry. Phosphorus metal and phosphoric acid from phosphorus are both also included with inorganic chemicals.

21. Data on materials handled are from Bureau of Mines (1989, Volume 1, tables 10, 11). Other data in this section on metals and minerals come from individual chapters in this same source.

22. The quantity of refuse produced obviously depends on the intensity of the beneficiation (washing) process. For comparison, the only coal cleaning process described in an official report of the United States Department of Energy (1980) had only a 70 percent yield in mass terms and a 90 percent yield in energy terms. Specifically, 1,428 tons of "run-of-mine" coal produced 1,000 tons of "clean" coal.

23· All data in this section are extracted from International Energy Agency (1991, pp. 664–665).

24. Light ends are compounds with boiling points in the range of butane (about 0 °C) and below. Methane and the light alkanes (C_2–C_4) fall into this category.

25. Benzene, ethylbenzene, toluene, and xylenes constitute, respectively, 0.1, 0.51, 0.19, and 0.88 percent of average crude oil by volume (Gaines and Wolsky, 1981). They constitute, of course, a much higher percentage of the volatiles.

26· Unless otherwise specified, the basic data for this section are from Bureau of Mines (1989).

27. Phosphorus pentoxide dissolved in water is phosphoric acid, the active ingredient in most phosphate fertilizers (e.g., superphosphates). It is not used, generally, in pure form.

28. The electrolytic process for chlorine production from brine yields 1.1 units of sodium hydroxide per unit of chlorine, with inputs of 1.75 units of sodium chloride. However, some chlorine is produced from magnesium chloride, and some is regenerated from hydrochloric acid, so the ratios are not exact.

29. Sodium chlorate used to bleach paper pulp is almost unique among chlorine chemicals. It is not manufactured from elemental chlorine but is made directly from sodium chloride (salt).

30. Feedstock data are from International Energy Agency (1991); consumption data for sulfuric acid, ammonia, fertilizer chemicals, and sodium carbonate (soda ash) are from Bureau of Mines (1988, 1989); data on shipments of inorganic chemicals from are *Chemical and Engineering News* (1997); and data on production and shipments of synthetic organic chemicals are from either *Chemical and Engineering News* (1997) or from United States International Trade Commission (ITC) (1989). The annual ITC reports, formerly a valuable source of data, unfortunately ceased publication in 1993.

31. Carbon black is used mainly in tires, of which it constitutes roughly 29 percent by weight (Ecoplan International, 1992). Total tire production in the United States in 1988 was roughly 2.2 MMT, accounting for 0.64 MMT of carbon, or 0.7 MMT of hydrocarbon feeds. There are other significant uses of carbon black, such as printing ink. However, most carbon black is made directly from natural gas. We do not include it as a chemical product.

32. Note that soluble cellulose is used to manufacture viscose rayon, cellulose acetate, and cellophane, among other products. Production in 1988 was 1.24 MMT, of which about 60 percent was used for rayon. Rayon is not counted as a product of the organic chemical industry.

33. Although about 0.2 MMT of phosphorus was used in detergents, most of it was inorganic: STPP and tetrasodium pyrophosphate.

34. Taking account of the availability of small amounts of other acids (HCl, HF, HBr, HNO_3, P_2O_5), it might seem that the need for sulfuric acid would be reduced somewhat. However, to the extent that other acids were used, the neutralization products would be sodium or ammonium salts of chlorine, fluorine, etc. Since these elements are actually embodied in products, they cannot also be a major constituent of the waste stream.

35. Unless otherwise specified, data for this section are from Bureau of Mines (1989, 1991).

36. Assuming that the iron in ore is mostly in the form Fe_2O_3, the 57.5 MMT of iron content in ore (1988) would be combined with 25.55 MMT of oxygen.

37. Unless otherwise specified, data for this section were taken from United States Bureau of Mines (1989).

38. Unless otherwise specified, data for this section were taken from United States Bureau of Mines (1989).

39. U.S. coal is unusually low in ash, most of which is recovered. By contrast, most other countries burn coal that has a much higher ash content, 15–25 percent or more, very little of which is recovered. Therefore, the problem of heavy-metal pollution from coal burning will be far more serious in eastern Europe, the former Soviet Union, China, and India.

REFERENCES

Brown, L. R., and E. C. Wolf. 1984. Soil Erosion: Quiet Crisis in the World Economy. Worldwatch Paper (60). Washington D.C.: Worldwatch Institute.

Bureau of Mines. 1987. Minerals Yearbook. Washington, D.C.: U.S. Government Printing Office.

Bureau of Mines. 1988. Minerals Yearbook. Washington, D.C.: U.S. Government Printing Office.

Bureau of Mines. 1989. Minerals Yearbook. Washington, D.C.: U.S. Government Printing Office.

Bureau of Mines. 1991. Minerals Yearbook. Washington, D.C.: U.S. Government Printing Office.

Bureau of the Census. 1983. Census of Manufactures. Washington, D.C.: U.S. Government Printing Office.

Bureau of the Census. 1988. Statistical Abstract of the United States: 1988, 108 ed. Washington, D.C.: U.S. Government Printing Office.

Bureau of the Census. 1990. Statistical Abstract of the United States: 1990, 110 ed. Washington, D.C.: U.S. Government Printing Office.

Bureau of the Census. 1991. Statistical Abstract of the United States: 1991, 111 ed. Washington, D.C.: U.S. Government Printing Office.

Carbon Dioxide Information Analysis Center. 1990. Trends '90: A Compendium of Data on Global Change. Oak Ridge, Tenn.: Oak Ridge National Laboratory.

Chemical and Engineering News. 1992. Chemical earnings. Volume 70.

Chemical and Engineering News. 1997. Chemical earnings: Japanese chemical producers generally fared well in 1996, but petrochemicals ate into profits. Volume 75.

Crutzen, P. J. 1976. The nitrogen cycle and stratospheric ozone. Paper presented at the Nitrogen Research Review Conference, National Academy of Sciences, Fort Collins, Colo., October 12–13.

Crutzen, P. J., I. Aselman, and W. Seiler. 1986. Methane production by domestic animals, wild ruminants, other herbivorous fauna, and humans. Tellus 38B:271–284.

Deevey, E. C. 1970. Mineral cycles. Scientific American 223:148–158.

Ecoplan International. 1992. New Developments in Tire Technology: Technological Change and Intermaterials Competition in the 90s and Beyond. Multi-Client Industry Report. Paris: Ecoplan International.

Gaines, L. L., and A. M. Wolsky. 1981. Energy and Materials Flows in Petroleum Refining. Technical Report (ANL/CNSV-10). Argonne, Ill.: Argonne National Laboratory.

Hoffman, M. S., ed. 1991. The World Almanac and Book of Facts. New York: Scripps-Howard.

Holmbom, B. 1991. Chlorine Bleaching of Pulp: Technology and Chemistry, Environmental and Health Effects, Regulations and Communication. Case study. April. Turku/Abo, Finland: Abo Akademi.

International Bank for Reconstruction and Development (IBRD). 1980. Environmental Considerations in the Pulp and Paper Industry. Washington, D.C.: IBRD.

International Energy Agency. 1991. Energy Statistics of OECD Countries 1980–1989. Paris: Organization for Economic Cooperation and Development.

International Trade Commission. 1992. Synthetic Organic Chemicals 1992. Washington, D.C.: U.S. Government Printing Office.

LeBel, P. G. 1982. Energy Economics and Technology. Baltimore, Md.: The Johns Hopkins University Press.

Lowenheim, F. A., and M. K. Moran. 1975. Faith, Keyes, and Clark's "Industrial Chemicals," 4th ed. New York: Wiley-Interscience.

Manzone, R. 1993. PVC: Life Cycle and Perspectives. Urbino, Italy: Commett Advanced Course.

Obernberger, I. 1994. Characterization and utilization of wood ashes. Technical paper. Institute of Chemical Engineering, Technical University, Graz, Austria.

Roskill Ltd. 1991. Chromium. London: Roskill Ltd.

Schlesinger, W. H. 1991. Biogeochemistry: An Analysis of Global Change. New York: Academic Press.

Schlesinger, W.H., and A.E. Hartley. 1992. A global budget for atmospheric NH_3. Biogeochemistry 15:191–211.

Science Applications International Corporation (SAIC). 1985. Summary of Data on Industrial Nonhazardous Waste Disposal Practices. Washington, D.C.: SAIC.

Smil, V. 1993. Nutrient flows in agriculture: Notes on the complexity of biogeochemical cycles and tools for modeling nitrogen and phosphorus flows in agroecosystems. Undated working paper.

Subak, S., P. Raskin, and D. von Hippel. 1992. National Greenhouse Gas Accounts: Current Anthropogenic Sources and Sinks. Boston, Mass.: Stockholm Environmental Institute.

Torrey, S., ed. 1978. Coal Ash Utilization: Fly Ash, Bottom Ash and Slag. Pollution Technology Review Series, No. 48. Park Ridge, N.J.: Noyes Data Corp.

Ulrich, A. H. 1990. U.S. Timber Production, Trade, Consumption and Price Statistics 1960–88. Miscellaneous Publication (1486). U.S. Forest Service. Washington, D.C.: U.S. Forest Service, U.S. Department of Agriculture.

United Nations Statistical Office. 1988. Industrial Statistics Yearbook: Commodity Production Statistics 1988 II. New York: United Nations.

United States Congress, Office of Technology Assessment. 1984. Wood Use: U.S. Competitiveness and Technology, Vol. II–Technical Report. OTA-M-224. Washington, D.C.: Office of Technology Assessment.

United States Congress, Office of Technology Assessment. 1992. Managing industrial solid wastes from manufacturing, mining, oil and gas production, and utility coal combustion. Background paper (OTA-BP-O-82). Washington, D.C.: Office of Technology Assessment.

United States Department of Agriculture. 1990. Agricultural Statistics. Washington, D.C.: U.S. Government Printing Office.

United States Department of Agriculture. 1991. Agricultural Statistics. Washington, D.C.: U.S. Government Printing Office.

United States Department of Agriculture. 1992. Agricultural Statistics. Washington, D.C.: U.S. Government Printing Office.

United States Department of Commerce. 1992. Current Industrial Reports: Inorganic Chemicals. Washington, D.C.: U.S. Department of Commerce.

United States Department of Energy. 1980. Technology Characterizations. Technical Report DOE/EV-0061/1. Washington, D.C.: U.S. Department of Energy.

United States Environmental Protection Agency, Office of Solid Waste. 1986. Waste Minimization Issues and Options. NTIS PB-87-114369 (EPA-530-SW-86-041). Washington, D.C.: U.S. Environmental Protection Agency.

United States Environmental Protection Agency. 1988. Solid Waste Disposal in the United States. Report to Congress (EPA-530-SW-88-011). Washington, D.C.: U.S. Environmental Protection Agency.

United States Environmental Protection Agency, Office of Solid Waste. 1991. 1987 National Biennial Report of Hazardous Waste Treatment, Storage and Disposal Facilities Regulated Under RCRA. NTIS PB-87-114369 (EPA-530-SW-91-061). Washington, D.C.: U.S. Environmental Protection Agency.

Measures of Environmental Performance and
Ecosystem Condition. 1999. Pp. 157–174.
Washington, DC: National Academy Press.

National Material Metrics for Industrial Ecology*

IDDO K. WERNICK AND JESSE H. AUSUBEL

Industrial ecology studies the totality of material relations among different industries, their products, and the environment. Applications of industrial ecology should prevent pollution, reduce waste, and encourage reuse and recycling of materials. By displaying trends, scales, and relations of materials consumed, emitted, dissipated, and discarded, metrics can expose opportunities to improve the performance of industrial ecosystems.

Metrics can indicate environmental performance at all levels: factory, firm, sector, nation, and globe. National metrics focus attention on collective behavior, particularly in a large country such as the United States whose economy sums the actions of more than 250 million people and 3 million for-profit corporations. The federal government assembles national data on a vast array of activities. The need is for a coherent set of metrics that enables efficient diagnosis of national environmental conditions and provides help in considering strategies for the future.

The need to develop environmental metrics is particularly strong for materials. National materials consumption indicates the structure of national industrial activity and its extent. Environmentally important industries such as mining, forestry, agriculture, construction, and energy production can be evaluated based on their material requirements and outputs. Despite their ubiquity and close association with environmental quality, materials have received little systematic analysis, particularly as compared with energy. This inattention stems in part from the heterogeneity of materials used in the modern economy and the myriad enterprises involved in transforming, processing, and disposing of materials and goods.

---

*A version of this paper appeared first in Vol. 21, No. 3 of *Resources Policy*. © 1995 Elsevier Science Ltd., Oxford, England. Reprinted by permission.

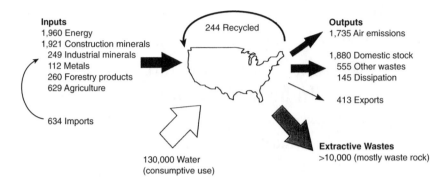

Inputs
1,960 Energy
1,921 Construction minerals
249 Industrial minerals
112 Metals
260 Forestry products
629 Agriculture

634 Imports

244 Recycled

130,000 Water
(consumptive use)

Outputs
1,735 Air emissions

1,880 Domestic stock
555 Other wastes
145 Dissipation

413 Exports

Extractive Wastes
>10,000 (mostly waste rock)

FIGURE 1 U.S. materials flows, circa 1990. All values are in million metric tons per year. Consumptive water use is defined as water that has been evaporated, transpired, or incorporated into products and plant or animal tissue and is therefore unavailable for immediate reuse. For a detailed description of this figure and data sources see Wernick and Ausubel (1995).

With the help of the Bureau of Mines, we have developed an environmentally oriented framework for characterizing material flows in the United States.[1] Choosing metrics requires a grasp of the diversity and enormity of U.S. materials flows (Figure 1). Our framework considers primarily three components: inputs to the economy (including imports), outputs (including exports), and extractive wastes. We aim for comprehensiveness in this framework in the sense that we do not want to "lose" materials and would eventually hope to record the complete materials balance. Our choice of inputs and outputs as major categories derives from the simplest of materials-flow models. We group extractive wastes separately because they represent immense mobilizations of materials readily distinguished from commodities, products, and other wastes. We use previously published data for all the values indicated and generally adhere to existing classifications.

We segment inputs into energy, construction minerals, industrial minerals, metals, forestry products, and agricultural products. We class outputs as domestic stock,[2] atmospheric emissions, other wastes, dissipation, and recycled materials. Imports and exports represent the masses of major individual commodities and classes of commodities crossing U.S. borders. Extractive wastes include residues from the mining and oil and gas industries. We account for water in Figure 1 but not in the material metrics because the weight and omnipresence of this resource would obscure what remains. We also omit consumption of atmospheric oxygen for biological respiration and in industrial processes.[3] We do not explicitly consider manufactured chemical products, but do include the mass of feed stocks used for organic and inorganic chemical production.

Materials have the advantage of offering a single unit of measure, weight, that allows for direct comparison across a broad range of material types. Kilograms and tons can hide variables such as volume, land disturbance, toxicity,[4] and other environmentally important qualities associated with materials that weight measures do not reflect. Nevertheless, weight does provide a reasonable starting point for appreciating the structure and scale of major activities affecting national environmental quality.

National material metrics do not obviate the need for monitoring environmental variables locally. Rather, they complement smaller-scale metrics that underscore the spatial distribution of problems and needs. In this respect, they resemble national economic indicators, such as gross domestic product (GDP). In addition, national materials metrics offer the prospect of capturing environmentally significant trends and relations not captured in the current regulatory framework, which tends to emphasize reporting by media, especially air and water, rather than along the functioning of the economic system.

NATIONAL MATERIAL METRICS

We propose eight general classes of metrics to indicate the current status and salient trends in national materials use as they influence environmental performance (Table 1). Most address either the productivity or the efficiency of resource use. Others indicate trends in the size and composition of materials use. Some metrics offer a means for quantifying aggregate environmental changes resulting from current national activities. Although some of the metrics are novel, others are already employed but gain meaning from the more systematic context. Although imperfect, this initial classification is intended to stimulate subsequent inquiry into the development of material metrics and the logic sustaining them.

Absolute National and Per Capita Inputs

The total mass of materials consumed by a nation, or individual members of its population, offers an indicator that tangibly values resource use. The components of the total differ in kind (and often in the accuracy of the supporting data), but their sum provides a benchmark for environmental management.

In 1990, each American mobilized on average about 20 metric tons of materials, or over 50 kg/day. The breakdown in Figure 2 equates with Figure 1 on national flows at the level of the individual American. This sum may be similar in other industrial nations. For example, estimates of Japanese materials use in 1990 total 52 kg per capita per day, a number closely comparable to the U.S. estimate (Gotoh, 1997).

The dynamics of per capita resource use as well as the efficacy of various policy initiatives aimed at affecting it could be gauged by comparing this number over time and across nations. More detailed metrics would look at consumption of classes of materials, such as energy fuels or agricultural minerals, and environmentally significant individual materials, such as lead.

TABLE 1 National Material Metrics

	Metric	Dimensions	Formula	Environmental Significance
Total per capita inputs		Metric tons/Capita	Aggregate consumption of all materials classes and individual material classes per capita	Benchmarking national resource use
Input composition	Fuel ratio	Dimensionless	Consumption ratio for coal:oil:natural gas	CO_2 emissions, cleanliness of the energy system
	Nonrenewable organics ratio	Dimensionless	Consumption quantity for nonrenewable organics/total hydrocarbon consumption	Petrochemical pollution, character of solid waste
	Structural materials ratio	Dimensionless	Consumption ratio of metals, ceramics, and polymers in all finished products and structures	Gross shifts in materials use, materials efficiency and cyclicity, mining and processing waste, energy use
	Agricultural ratios	Dimensionless	Consumption ratio of food to feed crops, rice and legume ratio to total agricultural produce	Land use, methane emissions, nitrogen fixation rates
Input intensities	Intensity of use	Million metric tons (MMT) of inputs/10^6 GDP	Material consumption quantity for selected input materials/GDP in constant dollars	Relationship of resource use to economic activity
	Agricultural intensity	Dimensionless	Fertilizer, pesticide, and agricultural minerals consumption/Total crop production	Materials efficiency, eutrophication of water bodies, topsoil erosion, chemical dissipation
	Decarbonization	MMT of carbon inputs/10^6 GDP	Carbon inputs/GDP in constant dollars	Relationship of carbon emissions to economic activity
Recycling indices	"Virginity" index	Percentage	Consumption of all virgin materials/Total material inputs	Materials efficiency and cyclicity, mining and processing waste, energy use

	Metals recycling rate	Percentage	Quantity of recycled and secondary metals consumption/Total metals consumption	Materials efficiency and cyclicity
	Renewable net carbon balance	Percentage	Forest growth/Forest products harvested	Global carbon balance of sources and sinks, land use, ecosystem disruption
Output intensities	Green productivity	Percentage	Quantity of solid wastes/Quantity of total solid physical outputs	Materials efficiency and cyclicity
	Intensity of use for residues	MMT/$\$10^6$ GDP	Generation quantity for selected materials waste streams/GDP in constant dollars	Relationship of waste generation to economic activity
Leak indices	Dissipation index	Percentage	Quantity of materials dissipated into the environment/Total material outputs	Materials efficiency and cyclicity, media contamination
	Nutrient and metals loadings	mg/liter and kg/km^2	Concentrations of pollutants in water bodies, and land deposition of nutrients and heavy metals/Defined area	Materials monitoring and accounting, media contamination
Environmental trade index		MMT	Net mass value of waste and emissions generated from foreign trade in manufactured products and raw resources	Domestic resource consumption, domestic environmental burden caused by exported goods
Mining efficiency	Mining wastes	Dimensionless	Quantity of wastes generated/Ton of finished product	Solid wastes, acid mine drainage
	By-product recovery	Dimensionless	Total by-product recovery/Total output	Materials efficiency, solid wastes

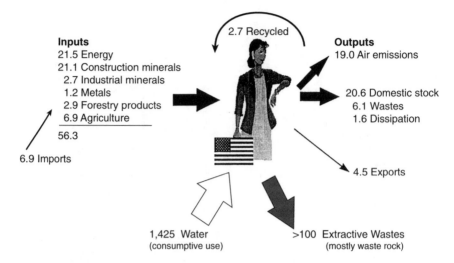

Inputs
21.5 Energy
21.1 Construction minerals
2.7 Industrial minerals
1.2 Metals
2.9 Forestry products
6.9 Agriculture
——————
56.3

6.9 Imports

2.7 Recycled

Outputs
19.0 Air emissions

20.6 Domestic stock
6.1 Wastes
1.6 Dissipation

4.5 Exports

1,425 Water
(consumptive use)

>100 Extractive Wastes
(mostly waste rock)

FIGURE 2   Per capita material flows, United States, circa 1990. All values are in kilograms per day. See caption for Figure 1 for further explanation.

Composition of Material Inputs to the National Economy

With economic development and technical change, the demand for materials evolves. Input composition reveals economic structure and dynamics and helps anticipate environmental consequences.

For example, environmental import attaches to the evolving ratio of the three fossil fuels used for energy, coal, oil, and gas, or in more elemental terms to the balance of hydrogen and carbon used to power and heat the nation (Marchetti, 1989; Nakicenovic, 1996). Although not used for energy, nonrenewable organic materials derived from petroleum and natural gas such as petrochemicals, plastics, asphalt, fibers, and lubricants comprise an appreciable fraction, about 6 percent, of total hydrocarbon consumption (Bureau of Mines, 1991a). The endpoints for these materials matter environmentally and as such merit their own distinct measure as a fraction of all hydrocarbon consumption.

The choice of structural materials indicates trends relevant to national environmental performance as well. Demand for properties in industrial and consumer goods influences selection among the major classes of structural materials: metals, ceramics and glasses, and polymeric materials including wood (Ashby, 1979). These materials range widely in their ability to bear loads, resist fracture, and operate in harsh thermal conditions. They also differ in typical densities (Figure 3). Similarly, they possess varying environmental attributes such as the energy needed, waste generated, and toxins released to the environment during extraction and processing. Comparing the energy needs for processing an equal

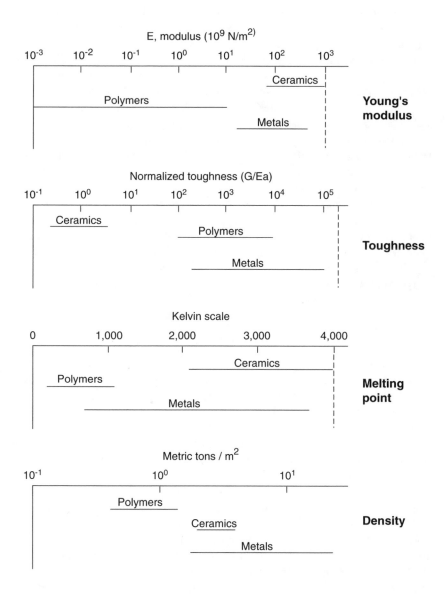

FIGURE 3 Range of physical properties for structural materials. Young's modulus is a measure of material elasticity. Toughness is a measure of resistance to fracture. Toughness is measured in units of joules per square meter of fracture surface (G) and is here normalized to Young's modulus (E) times atomic size (a). SOURCE: After Ashby (1979). Other sources include Carter and Paul (1991) and Hodgman (1962).

mass of aluminum, steel, cement, and polystyrene yields an approximate ratio of 85:10:2:1 (Agarwal, 1990; Hocking, 1991). Of course, materials rarely substitute for one another in products in a 1:1 mass ratio.

Historically, substantial scientific and engineering effort has been directed at improving the properties of metal alloys. Future gains may come in the area of polymers stiffened in the direction of loading, ceramics toughened to resist fracture, and composite materials designed to accentuate the best qualities (i.e., light, strong, and tough) of each material class. Although advanced materials may be difficult to reprocess, recyclability is not the single measure of environmental friendliness. This property must be weighed against gains derived from shifting to materials that perform functions using less mass, require less energy to process, and generate less incidental waste.

The composition of the food we consume, directly or indirectly, impacts the environment. Reduced national meat consumption accompanied by a rise in fruit, grain, and vegetable consumption diminishes the acreage used for grazing and feed in favor of less land-extensive crops. Cultivation of legumes and rice affects nitrogen fixation rates and atmospheric methane concentrations, respectively. Fertilizer and pesticide use rates are tailored to specific crops. In this case as with the others, input composition metrics clarify the environmental dimension of varying the mix of materials society consumes and shed light on paths for future development.

Intensities of Use

Intensity-of-use metrics show the evolution of individual materials used in the national economy by indexing primary, as well as finished, materials to GDP (Figure 4; also, see Malenbaum, 1978). These measures inform policy choices relating to natural resources by helping to gauge developmental status and to define realistic goals that integrate economic growth and improved environmental quality. In the energy sector, the declining intensity of carbon use, "decarbonization," of the U.S. economy relative to economic activity as well as energy use has been well established (Figure 5).

Intensity-of-use metrics also can show physical resource efficiency. For example, in 1990, the ratio of agricultural produce (e.g., grain, hay, fruit, and vegetables) to fertilizer inputs (e.g., nitrogen compounds and phosphates) was roughly 10:1 (Bureau of Mines, 1991b; United States Department of Agriculture, 1992). The ratio of food actually consumed by humans to mineral inputs is considerably lower. Other sectors using raw inputs as well as auxilliary materials for production (e.g., iron ore, coke, and lime for steel; wood and chemicals for paper) could apply similar environmental performance measures.

"Virginity" and Recycling Indices

A virginity, or raw materials, index measures the ratio of national raw materials use to total national inputs. It monitors the distance a society must go to stop

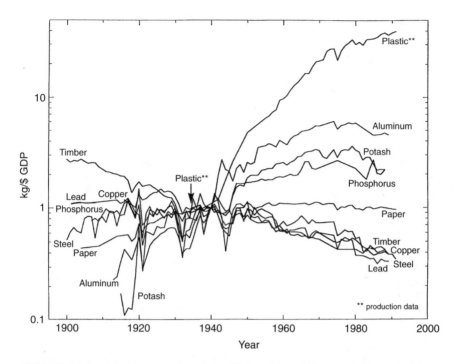

FIGURE 4 Materials intensity of use in the United States, 1900–1990. This metric conveys the evolving materials requirements of an economy over time. Consumption data are indexed to annual GDP in constant 1982 dollars. (For example, in 1900, U.S. phosphate consumption was 1,515,425 metric tons and gross national product was $261.5 billion, equivalent to about 5.8 metric tons per million dollars GDP. In 1990, 4,692,919 metric tons of phosphate were consumed and GDP was $4,120 billion, equivalent to about 11.2 metric tons per million dollars GDP.) All intensity-of-use values are normalized to unity at 1940 with the exception of plastics, which is indexed to 1942. SOURCES: *Modern Plastics Magazine* (1960); Bureau of the Census (1975, 1992). Data on U.S. production of plastics resin are from Broyhill, Statistics Department, Society of the Plastics Industry, Washington, D.C., personal communication, August 20, 1993.

extracting materials from the earth and sustain itself through its above-ground materials endowment and recycling. For 1990, recycled material accounted for about 5 percent of all inputs to the U.S. economy by weight (Rogich, 1993). Impeding the increase of this fraction are the heterogeneity of materials in the waste stream, industrial demand for materials with highly specific properties, and cumbersome regulations. These factors combine to shrink the pool of resources that can be used as inputs to production (Frosch, 1994; Wernick, 1994).

Among specific materials of interest are metals and wood. The fraction of secondary to total metals consumption indicates both the efficiency of metals

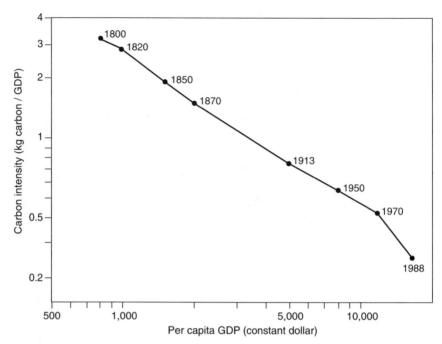

FIGURE 5 Diminishing carbon intensity of per capita GDP in the United States, 1800–1988. Carbon intensity is carbon consumed for energy divided by annual GDP in constant 1985 dollars. SOURCE: After Gruebler and Fujii (1991).

reuse from new scrap generated within industry and the success in recycling old scrap recovered from obsolete products such as automobiles. Recycling today accounts for over half the metals consumed in the United States (Figure 6; Rogich, 1993). However, recovery remains below 10 percent for arsenic, barium, chromium, and other biologically harmful metals listed in the Toxic Release Inventory (Allen and Behmanesh, 1994). The difference between annual forest growth and removal of growing stocks offers a simple measure of incremental changes in forest volume.[5] For the period 1970–1991, U.S. forests gained an average of over 150 million cubic meters of timber annually, augmenting existing timber volume at an annual rate of about 0.7 percent (United States Department of Agriculture, 1992).

Waste (Emission) Intensities

Waste intensities measure residuals and emissions per unit of output in physical or economic terms. Corporate practice increasingly evaluates the ratio of wastes to total firm output, including products and salable by-products (3M Corporation, 1991) and seeks uses for wastes (Ahmed, 1993; Edwards, 1993) as efficiency measures. National indicators would assess "green" productivity by evaluating the amount of materials considered as waste against various output categories. Figure 7 shows long-term trends of U.S. municipal solid-waste (MSW) generation, sulfur dioxide emissions, and emissions of nitrogen oxides indexed to economic activity. Industrial wastes are strong candidates for analysis using this metric. However, dry weight data on industrial wastes rarely exist or are hard to obtain (United States Congress, Office of Technology Assessment, 1992).

Leak Indices

Leak indices measure the ratio of outputs emitted and dissipated to total outputs, thereby quantifying the proportion of materials lost to further productive use and dispersed into the environment. Applying this measure allows for easier identification and isolation of "holes" in the system and focuses efforts to plug them.

Geographical information on nutrient and heavy-metals loadings aids improvement of accounts of dissipated materials. National efforts in this area are

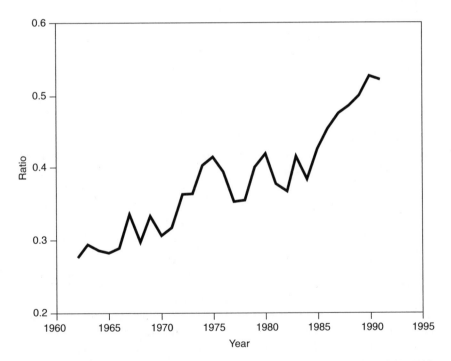

FIGURE 6 Ratio of secondary to primary metal consumption, United States, 1962–1991. SOURCE: Rogich (1993).

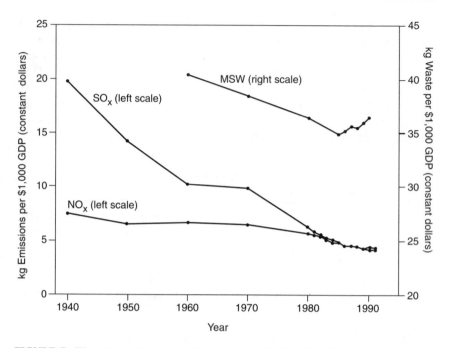

FIGURE 7 Waste intensities in the United States, 1940–1990. Municipal solid-waste (MSW) discards, and sulfur dioxide and nitrogen oxide emissions, indexed to GDP in constant 1987 dollars. SOURCES: Bureau of the Census (1975, 1994).

well established but incomplete. The National Oceanic and Atmospheric Administration (1993) estimates coastal discharges of nutrients (nitrogen, phosphorus), heavy metals (e.g., arsenic, lead, cadmium), and petroleum hydrocarbons in U.S. estuaries in the National Coastal Pollution Discharges Inventory. Estimates of inland nutrient discharges and metals deposition rates are sparse at best. Extending these measures to the entire nation would be laborious but worthwhile from the perspective of national environmental management.

Environmental Trade Index

An environmental trade index indicates the degree to which the nation is retaining or displacing pollution through international trade. Exporting raw materials consumes national resources and scars the domestic landscape. Using domestic industry to convert imported materials into finished goods and prepare indigenous materials for export can damage the environment in other ways. Despite intense interest in the monetary balance of U.S. foreign trade, the environmental profile of trade flows has received scant attention until recently, in the context of trade with Mexico.

By weight, commodities dominate trade. The mass of manufactured products traded contributes little to the total but may be responsible for domestic waste generation and discharges to the environment. During 1990, exports were dominated by agricultural products (33 percent), coal (23 percent), and chemicals (10 percent), all goods associated with domestic pollution. In the same year, crude oil and petroleum products accounted for over 60 percent of U.S. imports by weight, with metals and minerals accounting for another 20 percent (Bureau of the Census, 1993). We lack ready means to assess how the spatial redistribution of economic functions would affect environmental quality.

Extractive Waste Ratios

Extractive waste ratios measure resource efficiency in the mining industry. Recalling Figure 1 confirms the massiveness of wastes generated in this sector. Rock removed to expose mineral and ore bodies accounts for most of this waste. This material may be harmless, but exposing raw earth to wind and water can raise local acidity levels and allows for transport of trace elements. The sheer amounts of materials mobilized in mining and the economic incentive to minimize wastes combine with environmental objectives to advocate metrics of efficiency. Geological characteristics primarily determine overburden and tailings generated, but judgmental variables also affect mine wastes. One measure, subject to some physical constraints, is the amount of mine wastes per ton of mineral or ore mined, or primary metal produced. A separate useful measure, already used at the company level, looks at other inputs such as water and energy use per ton of finished product (Chiaro and Joklik, 1997). Measures of the recovery of by-products (e.g., methane in coal seams, sulfuric acid from smelter emissions, and metals from flue dusts) provide further examples of environmental indicators for the mining and mineral processing sector.

DISCUSSION

Industry operates and people behave within a system that evolves to satisfy human wants and uses a dynamic set of means to achieve them. As a discipline, industrial ecology discourages reducing the system to components and examining them in strict isolation. The challenge for national material metrics, as well as other national environmental metrics, is to quantify and integrate relevant data that elucidate the primary structure and development of the system from an environmental perspective.

National material metrics rely on empirical data. Various agencies of the federal government collect relevant data for one purpose or another. However, unless coordinated, the data do not fully support existing metrics and limit the scope for future ones. Procedural changes aimed at synchronizing data collection among various federal departments and agencies to build a single base (year) would amplify the benefits of existing collection efforts. Equally important from an environmental perspective is the development of standardized definitions for classifying material commodities to erase confusion leading to omissions and double counting of material components.

Accurate data on wastes are the hardest to obtain. Companies collect little or no data for many waste streams due to the actual or perceived absence of economic value. High disposal costs and regulatory requirements have improved waste accounting practices at many firms, but wastes have yet to receive the respect that marketability confers. Among the main goals of industrial ecology is exploring potential markets for waste materials. Currently, the dearth of reliable information available for wastes is one of the factors blocking progress. Better information would improve the market climate for wastes and at the same time help to develop metrics that assess their relative impact nationally.

Although improved national environmental metrics go hand in hand with better databases, metrics are not meant simply to compile information. Their purpose is to embed the data in a context that recognizes the larger system and is relevant to how it works. Good environmental indicators exist, but too often remain detached from each other and from an unambiguous framework. Appropriate metrics should correlate individual indicators and clarify the relation of each one to the whole. To illustrate, citing fertilizer usage rates without reference to agricultural productivity is misleading and causes unwarranted alarm. Conversely, extolling the environmental virtue of a lighter consumer product without examining the life-cycle implications of its fabrication and disposal is premature. To enhance their value and minimize misuse, commentary and interpretation should accompany the publication of metrics.

To adequately respond to complex questions of environmental performance requires both context and an array of metrics. For example, is the nation beginning to "dematerialize," that is, effectively decouple overall materials consumption from continued economic growth? For the U.S. energy sector the answer has been in the affirmative. Efficiency gains and the shift away from heavy manufacturing have modified the traditional relation between energy consumption and economic growth in the United States. Single indicators (i.e., kilowatt hours consumed/$GDP) elegantly illustrate this development. To have similar confidence regarding materials will require a more elaborate set of measures that are sensitive to the diverse structure of contemporary materials use and the many forces affecting its dynamics (Wernick et al., 1996). National materials metrics would refine how such questions are articulated and provide the basis for more convincing answers than are now available.

Looking to the future, national materials metrics help order the national research agenda for materials science and engineering (National Academy of Sciences, 1989). At over 50 kg per day per American, even the rough profile developed here demonstrates the need for meshing environmental and materials research. Metrics highlight the locations and relative urgency of incorporating environmental goals into materials research programs. Significantly, these goals often overlap with factors affecting the bottom line such as reducing inputs, improving efficiency, recycling, and complying with environmental regulations.

Future materials fluxes, including both products and by-products, may even exceed contemporary ones in size. To make them environmentally compatible, we need better methods for analyzing their current condition and anticipating future changes. To achieve the goal of a more circular economy, society needs to consider its materials legacy as a dowry to future generations, rich in valuable ore. By capitalizing on the "mines above ground" or scrap piles for materials, wastes from extraction and disposal grow dispensable. We can imagine an industrial ecosystem in which emissions, including carbon and water vapor, are captured and complex waste streams are separated to recover the value and utility of their components. The discipline of creating national materials metrics is a useful start to creating a consistent, realistic long-range technical vision.

ACKNOWLEDGMENTS

We are indebted to Donald Rogich, Jim Lemons, and Grecia Matos at the Bureau of Mines for data and ideas on materials taxonomy.

NOTES

1. In this paper we draw on other work by the authors (Wernick and Ausubel, 1995) that contains detailed data supporting the metrics presented here.
2. Domestic stock refers to materials embedded in structures and products not discarded for a period longer than 1 year.
3. We include atmospheric nitrogen fixed into NOx emissions as well as for ammonia production. We omit estimates of the mass of soil eroded during agricultural operations.
4. A clear example of this is annual total U.S. dioxin and furan emissions, which are counted in kilograms rather than tons, yet have considerable environmental impact (Thomas and Spiro, 1995).
5. A complete net carbon balance for forests includes annual carbon flows in trees, soil, forest floor, and understory vegetation. Since 1952, the amount of carbon stored in U.S. forests has grown 38 percent, adding about 9 billion metric tons of carbon (Birdsey et al., 1993).

REFERENCES

Agarwal, J. C. 1990. Minerals, energy, and the environment. Pp. 389–395 in Energy and the Environment in the 21st Century, J. W. Tester, D. O. Wood, and N. A. Ferrari, eds. Proceedings of a conference held at Massachusetts Institute of Technology, Cambridge, Mass., March 26–28, 1990.

Ahmed, I. 1993. Use of Waste Materials in Highway Construction. Park Ridge, N.J.: Noyes Data Corp.

Allen, D. T., and N. Behmanesh. 1994. Wastes as raw materials. Pp. 69–89 in The Greening of Industrial Ecosystems, B. R. Allenby and D. J. Richards, eds. Washington, D.C.: National Academy Press.

Ashby, M. F. 1979. The science of engineering materials. Pp. 19–48 in Science and Future Choice, Vol. I, P. W. Hemily and M. N. Özdas, eds. North Atlantic Treaty Organization. Oxford, England: Clarendon Press.

Birdsey, R. A., A. J. Plantinga, and L. S. Heath. 1993. Past and prospective carbon storage in United States forests. Forest Ecology and Management 58:33–40.

Bureau of Mines. 1991a. Minerals Yearbook 1991. Washington, D.C.: U.S. Government Printing Office.

Bureau of Mines. 1991b. Mineral Commodity Summaries. Washington, D.C.: Bureau of Mines.

Bureau of the Census. 1975. Historical Statistics of the United States, Colonial Times to 1970. Washington, D.C.: U.S. Government Printing Office.

Bureau of the Census. 1992. Statistical Abstract of the United States: 1992, 112th ed. Washington, D.C.: U.S. Government Printing Office.

Bureau of the Census. 1993. Statistical Abstract of the United States: 1993, 113th ed. Washington, D.C.: U.S. Government Printing Office.

Bureau of the Census. 1994. Statistical Abstract of the United States: 1994, 114th ed. Washington, D.C.: U.S. Government Printing Office.

Carter, G. F., and D. E. Paul. 1991. Materials Science and Engineering. Materials Park, Ohio: The Materials Information Society.

Chiaro, P., and F. Joklik. 1997. Environmental strategies in the mining industry: One company's experience. Pp. 165–181 in The Industrial Green Game: Implications for Environmental Design and Management, D. J. Richards, ed. Washington, D.C.: National Academy Press.

Edwards, G. H. 1993. Consumption of Glass Furnace Demolition Waste as Glass Raw Material. Paper presented at the National Academy of Engineering Workshop on Corporate Environmental Stewardship, Woods Hole, Mass., August 10, 1993.

Frosch, R. A. 1994. Industrial ecology: Minimizing the impact of industrial waste. Physics Today 47(11):63–68.

Gotoh, S. 1997. Japan's changing environmental policy, government initiatives, and industry responses. Pp. 234–252 in The Industrial Green Game: Implications for Environmental Design and Management, Washington, D.C.: National Academy Press.

Gruebler, A., and Y. Fujii. 1991. Inter-generational and spatial equity issues of carbon accounts. Energy 16(11/12):1397–1416.

Hocking, M. B. 1991. Paper vs. polystyrene: A complex choice. Science 251:504–505.

Hodgman, C. D. 1962. Handbook of Chemistry and Physics, 44th ed. Cleveland, Ohio: Chemical Rubber Publishing Company.

Malenbaum, W. 1978. World Demand for Raw Materials in 1985 and 2000. New York: McGraw-Hill.

Marchetti, C. 1989. How to solve the CO_2 problem without tears. International Journal of Hydrogen Energy 14(8):493–506.

Modern Plastics Magazine. 1960. Vol. 37, No. 5.

Nakicenovic, N. 1996. Freeing energy from carbon. Daedalus 125(3).

National Academy of Sciences. 1989. Materials Science and Engineering for the 1990s. Washington, D.C.: National Academy Press.

National Oceanic and Atmospheric Administration (NOAA). 1993. Point Source Methods Document. Strategic Environmental Assessments Division. Silver Spring, Md.: NOAA.

Rogich, D. G. (and staff). 1993. Materials Use, Economic Growth, and the Environment. Paper presented at the International Recycling Congress and REC'93 Trade Fair, Geneva, Switzerland. Washington, D.C.: Bureau of Mines.

Thomas, V. M., and T. J. Spiro. 1995. An estimation of dioxin emissions in the United States. Toxicological and Environmental Chemistry 50:1–37.

3M Corporation. 1991. Challenge '95: Waste Minimization. St. Paul, Minn.: 3M Corporation.

United States Congress, Office of Technology Assessment. 1992. Managing Industrial Solid Wastes from Manufacturing, Mining, Oil and Gas Production, and Utility Coal Combustion. OTA-BP-O-82. Washington, D.C.: U.S. Government Printing Office.

United States Department of Agriculture. 1992. Agricultural Statistics 1992. Washington, D.C.: U.S. Government Printing Office.

Wernick, I. K. 1994. Dematerialization and secondary materials recovery: A long run perspective. Journal of the Minerals, Metals, and Materials Society 46(4):39–42.

Wernick, I. K., and J. H. Ausubel. 1995. National materials flows and the environment. Annual Review of Energy and Environment 20:462–492.

Wernick, I. K., R. Herman, S. Govind, and J. H. Ausubel. 1996. Materialization and dematerialization: Measures and trends. Daedalus 125(3).

Information for Managers

Measures of Environmental Performance and
Ecosystem Condition. 1999. Pp. 177–187.
Washington, DC: National Academy Press.

Environmental Measures:
Developing an Environmental
Decision-Support Structure

REBECCA TODD

Given ever-present resource constraints, a substantial proportion of business managers' time and effort is devoted to maximizing output and profitability and minimizing risk and loss. In the increasingly competitive international markets where many firms operate, the choices about resource allocation have become ever more important to both the short term profitability and the long-term survival of the firm. In making choices, managers must continually trade off one feasible business opportunity against others, weighing the resources they expect to be consumed against the anticipated benefits for each. The ultimate quality of the decision made, that is, the relative benefits and costs achieved or realized, will depend heavily on the quality of information that the manager has available when the decision is made.

Information systems dedicated to supporting managers' decisions have long been an integral part of business. However, environmental decision making is relatively new as an intrinsic, fundamental part of ongoing business strategy development and definition. Consequently, the environmental information support that managers need to make decisions is rarely as well developed as the information available for other business areas. Most crucially, the quality of the available information may be poor. Moreover, upper-level managers, responsible for most strategy and policy formulation, may not be fully aware of the deficiencies of the information because details about the data's limitations are typically lost as the data are transmitted up through the firm's levels of management.

Few disciplines have been marked by such rapid change as those related to the environment. Many developing areas of corporate and academic research, for example industrial ecology and life-cycle analysis, were unknown 20 years ago, and today they consume significant resources. However, the relative infancy of these and other environmental efforts stymies the development of the measurement techniques and dedicated instrumentation needed to carry them out properly. This lag exists in part because firms face such an enormous task in removing waste from emissions and effluent streams as rapidly and fully as possible. These activities rightly have had to consume the lion's share of corporate environmental investment resources in the short term. As a result, however, most of the development has had to be directed to end-of-the-pipeline waste control rather than to prevention and control of input or process-stage pollution. As a consequence, we see relatively sophisticated measurement and control efforts being made over incineration and waste-water treatment processes, while little or no effort goes toward processes to handle the flows going into the facilities other than, perhaps, engineering estimates required for recharges of facilities' costs.

Previous studies have examined some of the qualitative environmental inadequacies of typical business accounting systems (Todd, 1989, 1994) as well as systematic approaches to environmental profiling and information gathering of firms' productive activities (see, for example, Allenby and Graedel, this volume; Allenby, 1994). This paper focuses on the development of a taxonomy for evaluating the adequacy of environmental measurement systems and information available within the firm, targeting those areas with the most urgent need for measurement development or refinement. The objective is to provide firms with a basic structure for (1) evaluating the current state of environmental measurement and control, and (2) developing a business strategy that will allow them to move systematically from this stage to a rigorous and comprehensive system of environmental monitoring, control, and managerial decision support. Other environmental information systems, for example Allenby and Graedel's (this volume) environmentally responsible facility assessment matrix, could prove invaluable as basic building blocks in the structure. Beyond this immediate objective, the ultimate goal is to develop a metasystem to support pollution prevention efforts at the source. This paper first examines the types of decisions that managers must make, especially about the environment. Next, it considers the nature of measurement and some of the factors that reduce the quality of the measures. Finally, it describes the development of an environmental measurement taxonomy and the types of managerial decisions that are likely to be affected.

MANAGERS' BUSINESS DECISIONS

Typically, managers must make decisions that affect three broad categories of operations: (1) long-term capital investment, (2) monitoring and control of operations, and (3) performance evaluation and motivation of employees. Environmental considerations are increasingly involved in each of these.

Among the most difficult decisions that managers must make are those that involve the long-term dedication of scarce capital to direct asset investment or research and development projects. Such investments are typically undertaken to improve profitability or competitive position or to reduce risk. The specific goals may be, for example, to develop new products or markets, increase production yields, or reduce wastes. More and more frequently in recent years, these goals have had environmental aspects, such as developing processes that eliminate toxic or highly reactive compounds or reduce air emissions, such as volatile organic compounds. With these investments, the manager may be able to achieve several major objectives at once: increase profitability, improve compliance with environmental regulations, and reduce the firm's long-term imbedded risk by decreasing or eliminating sources of future liabilities.

An obvious difficulty with such otherwise desirable investments is that they are likely to be highly uncertain undertakings and involve very long payback times, amounting to 10–15 years or more. Indeed, a middle-level manager proposing such long-term investments is unlikely to be there to reap the rewards of the investment under the performance evaluation systems that most firms use.

Therefore, firms that seek to minimize imbedded risk and improve competitive positions must provide senior managers responsible for long-term decision making with the best information obtainable, given cost and benefit constraints. This information must not only be relevant to the decision but also be as reliable as possible. Although large sums of capital may be expended in a single decision, only rarely do firms "invest" in major projects to improve the quality of information. Moreover, because environmental decision making is relatively new, firms may not be systematically gathering and aggregating information suitable for environmental decisions.

A second major use of managerial information is in the monitoring and control of ongoing operations. Decisions in this area are commonly regarded as being routine—for example, continuous monitoring of a chemical batch process, quality testing of products, and comparisons of actual and budgeted production output. However, managers responsible for this day-to-day routine monitoring are likely to be those with the greatest knowledge of specific processes and product markets, as well as with the environmental aspects of the operations. Therefore, they are in the best position to recognize and respond to changes in the environment, and also those of a more general business nature. Indeed, they may have most of the responsibility for achieving the goal of minimizing future environmental liabilities now. Therefore, the greater the precision, relevance, and validity of the measures they rely on, the likelier it is that they can take timely and effective action to minimize both short- and long-term environmental problems.

Performance evaluation and motivation of employees are among the most important uses of information for supporting managerial decisions. Most firms want to minimize waste and reduce environmental liability exposure. Generally, employees are encouraged to achieve corporate goals through the performance evaluation and rewards systems that are established specifically to accomplish this purpose. For example, because most companies want to increase profitability, many typical managerial reward systems award so-called incentive pay and bonuses to those managers who achieve planned profit targets. However, few companies have chosen to incorporate environmental goals into incentive compensation contracts. Indeed, although many firms stress environmental, health, and safety awareness and goals at all levels, the managers making the day-to-day decisions may view the environment as being of secondary importance at best. The reason is simple: The rewards system by which managers chart their own progress speaks only to profits, not pollution prevention or long-term risk reduction.

The difficulty of measuring environmental "progress" is a common reason cited for the omission of environmental concerns in the direct incentive and rewards scheme. Part of the difficulty is that both the risks and the paybacks in this area are uncertain and long term. Typical pollution-prevention projects in the past commonly have been characterized by easily measured inputs and outputs with immediately apparent cost-benefit trade-offs. Projects involving complex measurement difficulties are frequently ignored. However, an emphasis on short-term problems is not unique to the environmental area. Recently, both accounting regulatory organizations and firms have had to come to grips with such issues as the measurement of pension and postretirement (health care) benefits. Indeed, a major impetus for firms' shifting to health care cost-containment schemes is that, once they had developed workable measures of the future liabilities and the related current costs of these programs, managers were forced for the first time to monitor and control them.

Thus, gains to long-term strategic management of firms' competitive positions, profitability, and risk are likely to be enhanced substantially by investment of time and resources in the development of improved measures of environmental input, output, outcome, and performance.

THE NATURE OF MEASUREMENT

Measures, whether of the height and weight of a person or of emissions from volatile organic compounds, are likely to be flawed representations of the true natural state of the object being measured. The difference between the "true" natural state and that suggested by the measure, or proxy, is termed measurement error. The relationship between the true state and the measure is

$$x_{measured} = x_{true} + e_{systematic} + e_{random}$$

where $x_{measured}$ is the flawed proxy for what we wish to know, x_{true}; $e_{systematic}$ is a form of repetitive error; and e_{random} is a nonrepetitive and, by definition, unpredictable form of error. The measure of, for example, the amount of a certain toxic chemical in waste that is flowing to a water treatment facility, is the only directly observable component of the relationship. The systematic error, $e_{systematic}$, arises from sources of bias in the way the toxic chemical is measured. A simple example would be if a malfunctioning instrument were used in the analysis and caused the measurement to be consistently higher (or lower) than the actual amount of chemical present in the stream. Certainly, more than one source of systematic bias may be present in a given measure. e_{random} encompasses the remaining element of error. As the name suggests, this error source is unpredictable and nonrepeating. This portion is assumed to have a mean of zero over the long term, although individual observations may have large positive or negative random errors, depending on the nature of the measure.

The usefulness of a measure in decision making is directly related to its reliability. Reliability is the proportion of the measure that is free of error. For example, if 90 percent of a measure is either systematic or random error, the reliability is only 10 percent. Put differently, only 10 percent of the measure accurately reflects the object being measured. In many fields, for example medicine, much time and effort is expended on the development of highly reliable measures and tests. The reason is obvious: correct diagnosis and treatment and perhaps the patient's life depend on the accuracy of the informational inputs to the physician's decisions.

In business, however, the reliability of measures may vary widely. In traditional cost accounting, reasonably precise measures may be obtained for the direct materials and labor expended to manufacture a product. Other cost sources are measured with substantially less precision, including such overhead items as depreciation, the allocation of the historical cost of fixed assets to each of the periods in which the asset is used in production. Indeed, when precisely and imprecisely measured costs are summed, the result may not be useful for informing managerial decisions (Todd, 1994).

Such problems are likely to be particularly extreme in the environmental area. Several factors explain this, including the relative inexperience of firms' managers and financial staff in dealing with the much newer environmental measures, the high levels of complexity involved in much of the measurement, and uncertainty about the long-term environmental implications of many of the materials used in production.

As a consequence, many managers have come to rely on qualitative evaluations, rather than quantitative data, for environmental decision making. This certainly makes good sense where the knowledge base is low and measurement and monitoring technology is primitive or nonexistent. Indeed, the practice of medicine was in this state two centuries ago. However, if information determines the quality of both decisions and outcomes, it clearly is necessary to develop systems for evaluating and monitoring information quality and continually upgrading the information available to decision makers.

Common Sources of Environmental Measurement Error

Environmental measurement error may arise from a variety of sources. However, several circumstances account for a large proportion of such error.

In most cases, the most reliable environmental measurements that firms now use are those related to government environmental regulatory oversight. Given possible public audits by either regulators or public-interest groups, managers have incentives to ensure that the numbers they report are reasonably accurate. However, such measurement is largely centered on end-of-the pipeline emissions and effluents, rather than pollution prevention and elimination. Consequently, managerial decisions have been flavored to some extent by this regulatory and measurement focus. That is, a manager who must comply with an environmental discharge regulation will no doubt seek the most readily achievable fix, for example installing a scrubber to remove the emission. The difficulty is that the short-term regulatory solution may not be followed by long-term process-waste removal unless both reliable measures and incentives are in place to encourage managers to do so.

Managers are likely to use less-precise information for internal managerial environmental decisions than for reports required by regulations. For example, managers frequently rely on estimates of individual process waste streams and component volumes developed either in process research and development or in early stages of implementation. Sampling to confirm the accuracy of the estimates may occur infrequently or not at all. Without such confirmation, managers cannot know the accuracy or reliability of estimates. Moreover, if either processes or planned inputs are changed, the estimates may well not be updated. In addition, as facilities age, the yield and waste proportions may change.

Such situations may result in a very large systematic error component of waste-stream measures, which renders the measures unreliable and decisions based on the measures suboptimal or even dysfunctional. For example, waste treatment facilities' costs are typically redistributed by means of recharges (or transfer prices) to production units using the facilities' services, based on estimates of waste volumes and components. More important, however, such inaccuracies may strongly influence long-term investment decisions on capital asset replacement. Monitoring and control functions will fail if control targets are not known with reasonable accuracy.

Random measurement error results from short-term fluctuations in any of the factors that can influence the measure. These may include (but are not limited to) human error as well as external sources, for example inputs that do not meet engineering specifications. Unless the measurement is ongoing, such error may not be detected.

A TAXONOMY FOR IMPROVING
ENVIRONMENTAL MEASUREMENT

Current managerial information systems, including accounting systems, typically focus on only a relatively small subset of the data available or obtainable about a company's productive efforts. This occurs because these systems have been driven primarily by the external financial reporting imperative incumbent on all publicly held firms.

Regulatory bodies have defined both the type and the amount of financial information that firms are required to report. To minimize the cost of acquiring this information, firms have tended to adapt the information systems they use to support external reporting requirements. For example, external financial reporting rules require full-absorption (that is, fully aggregated) costs of goods sold and ending inventory valuation numbers. Full-absorption costs comingle not only current out-of-pocket expenses but also allocations of the historical costs of fixed assets.

Managers, rather than suffer additional delays while a new form of information is being specially processed or incurring additional costs of information development, will commonly use the readily available but, in many cases, highly unsuitable numbers for decision making. In the example just cited, a manager in the process of deciding whether, for the short term, to manufacture a product internally or buy it from an external vendor would want measures of the out-of-pocket, or short-term escapable, costs. All other factors are irrelevant. However, historical allocations built into the full-absorption cost may be so large as to swamp the out-of-pocket information, and it may not be possible to disentangle the irrelevant from the relevant decision information. In terms of the criteria for measurement quality in the foregoing section, a full-absorption number for a make-or-buy decision will likely contain a large proportion of systematic error that will be evident only after considerable investigation. If a manager wants to determine the proportion of a product's cost that has environmental implications, most information systems in use probably could not generate that information without considerable additional analysis.

In summary, existing managerial and accounting information systems may suffer from two broad classes of inadequacies related to environmental decision making: (1) the required information may not be available in the current system, and (2) the information available may be irrelevant or subject to substantial and indeterminate error as applied to the decisions to be made.

Therefore, for environmental decision-making purposes, including capital investment, monitoring and control, and performance evaluation and motivation, the firm must review or profile the environmental information generation, aggregation, and reporting that it uses. Table 1 suggests a simple taxonomy, readily adaptable to a wide variety of firms. Environmentally relevant functions or activities run across the top, indicated by numerals I to V. Depending on the nature of a firm's business and productive activities, the firm may have more or fewer such categories. However, to be comprehensive, the measurement profile should extend back to the relevant supplier/factor stage and forward to customer/end-use activities. This scheme is entirely consistent with life-cycle analysis approaches.

The second dimension of the taxonomy reflects the relative quality of the information available. Only four categories are suggested here, ranging from

TABLE 1 Taxonomy for Development of Information and Measurement Systems for Environmental Decision Making E

Measurement/ Information Quality	Environment-Related Function/Activity				
	I Suppliers	II Inputs/Processing	III Outputs	IV Waste Treatment/ Processing	V End Use/Recycling
A Precision measure	Materials, production methods, products, by-products, waste generated, waste handling, waste treatment, regulatory profile, legal liability issues etc...	Raw materials (nonenvironmental), raw materials (environmental), direct labor, relevant overhead, energy, ancillary services, production methods, regulatory profile, legal liability issues, etc...	Products (nonenvironmental), products (environmental), by-products (nonenvironmental), by-products (environmental), waste (nonenvironmental), waste (environmental), regulatory profile, legal liability issues, etc...	Waste handling, waste recycling, waste reprocessing, waste treatment (air), waste treatment (water), waste treatment (solid), ancillary services, regulatory profile, legal issues, etc...	Waste generated, waste recycling, waste reprocessing, waste treatment and disposal, regulatory profile, legal issues, etc...
B Quantitative measure					
C Qualitative evaluation					
D Measures not available					

necessary but currently nonexistent information to measures of the highest and most reliable quality. Again, this range can be adapted to meet a firm's needs in a particular case. For example, if a particular management decision is designed to identify which facilities are "environmentally responsible" (see Allenby and Graedel, this volume) and those that are substandard along one or more dimensions, qualitative, or category C, measures may be perfectly suitable. However, if research and development managers are trying to decide where to target scarce research and development resources according to which of a number of competing projects will yield the highest environmental benefit in terms of pollution reduction, then category B or even A measures may prove most useful.

The purpose of the taxonomy, then, is the early identification of information needs, sufficiencies, and inadequacies so that firms can undertake planned information system enhancements. In many cases, they can achieve the enhancements at very little additional cost, given adequate time and planning.

A handful of items that firms might consider are listed under the various activity categories. Traditional accounting systems, which concern themselves only with the productive activities within the firm, will not generate information about suppliers (category I). The one exception is vendor lists in the purchasing departments. More recently, however, many firms have begun to make more or less routine visits to suppliers to learn something about the suppliers' exposure to potential environmental liabilities, and thus the firm's possible liabilities as well. In such cases, they usually collect qualitative information. Nonetheless, the firm's environmental stewardship begins with the factor inputs and any business relationships that may be relevant.

Category II, inputs/processing, is usually the point where managerial accounting and information processes begin. For example, data on materials and supply inventories and on labor costs have long been accounted for relatively precisely. However, additional costs, usually designated overhead items, are normally pooled and reallocated to products rather than being explicitly accounted for. Energy costs, one of the most important costs for environmental measurement and control, are invariably treated in this way. Only rarely does the information system recognize the environmental relevance of such cost items. For example, raw materials may be toxic, highly reactive, or have other environmental implications, but only those with direct knowledge of the processes will have this information. Moreover, many items highly relevant to the environment, for example organic solvents, may disappear into overhead pools in the accounting process (Todd, 1994). Other category II costs, including ancillary services such as toxicity testing and legal advisory services, which may have high environmental relevance for a particular product, may not be linked to the product in any fashion. Indeed, the product-specific costs incurred will likely not be identified at all, becoming category D measures (or nonmeasures).

Beginning with the outputs category, specific internal accounting declines rapidly in quality insofar as environmental considerations are concerned. For example, because most managerial accounting and information systems focus on a salable manufactured product, detailed records on waste generated are unlikely to be available. Moreover, the system does not capture the distinction between environmentally relevant products, by-products, and wastes.

As observed above, information on waste recycling, reprocessing, and treatment activities is usually based on engineering estimates that may be out of date and substantially in error. Measures for category V, customer end-use and recycling activities, are at best in an early stage. Nonetheless, these are receiving increasing regulatory and consumer attention and are likely to become significantly more important in the future.

A latent but vital dimension in this taxonomy is the identification of pollution sources. For example, among the system's environmentally relevant outputs that may require waste recycling, reprocessing, or treatment may be a hazardous input that is only partially consumed in the process, thereby generating a hazardous waste. The resulting waste costs, including any incurred by regulatory monitoring and oversight, are commonly treated by accounting systems as if they arose in category IV. In fact, however, the problem originates in categories I and II and can only be eliminated by intervention there. Rather simple refinements to the information gathering and processing system will make it possible not only to identify the point sources of such items but to aggregate the entire effect of their use through the whole of the system.

CONCLUSIONS

This paper recommends that firms develop comprehensive systems for evaluating the adequacy of environmental information currently available to support managers' environmental decision-making needs and establish procedures to systematically upgrade those systems. The rationale for such a process is that the quality of managers' decisions is likely to be heavily influenced by the quality of information available to them.

A simple prototype taxonomy is presented that firms may adapt to their own operating environments and that can be used to guide the identification of relevant environmental information and the quality of information currently available in the firm. The firm can use this knowledge to facilitate the deployment of assets and the adoption of development efforts in those areas where information is deemed to be most severely lacking. The ultimate goal is to provide essential support for efforts to prevent pollution and reduce energy consumption across the entire spectrum of a firm's activities.

REFERENCES

Allenby, B. R. 1994. Integrating environment and technology: Design for environment. Pp. 137–148 in The Greening of Industrial Ecosystems, B. R. Allenby and D. J. Richards, eds. Washington, D.C.: National Academy Press.

Todd, R. B. 1989. Environmental Accounting: Patching the Environmental Fabric. National Research Council. Washington, D.C.: National Academy Press.

Todd, R. B. 1994. Zero-loss environmental accounting systems. Pp. 191–200 in The Greening of Industrial Ecosystems, B. R. Allenby and D. J. Richards, eds. Washington, D.C.: National Academy Press.

Bioassays

Measures of Environmental Performance and
Ecosystem Condition. 1999. Pp. 191–198.
Washington, DC: National Academy Press.

A Critique of Effluent Bioassays

CLYDE E. GOULDEN

The Clean Water Act of 1977 states, "It is the national policy that the discharge of toxic pollutants in toxic amounts be prohibited" (Peltier and Weber, 1985, p. 1). Thorough assurance that this goal is met would require complete chemical profiles of every effluent; knowledge of the sensitivity of all potentially affected organisms to all chemicals in effluents, including both direct toxic effects and indirect effects, such as the effects of toxins on forage species; and an understanding of all synergistic interactions between compounds in effluents. It is not feasible to obtain such comprehensive information.

The U.S. Environmental Protection Agency (EPA) concluded that a cost-effective alternative approach would be to measure effluent toxicity by exposing aquatic organisms to effluents in "bioassays." Bioassays measure "the potency of any stimulus, physical, chemical, or biological, physiological or psychological, by means of the reactions that it produces in living matter" (Finney, 1952a, p. 1). The rationale is that through bioassays, test organisms reveal whether an effluent is toxic. This paper describes the development of the bioassay approach and evaluates whether through its use ecosystems can be sufficiently protected from toxic materials.

THE HISTORY OF BIOASSAYS

The basic design of bioassays was developed in the nineteenth century, but test species were used as an assay of exposure to some stimulus well before that time. Finney (1952b) suggests that the basic principles of bioassays are found in early texts and quotes an example:

And it came to pass at the end of forty days, that Noah opened the windows of the ark which he had made:

And he sent forth a raven, which went to and fro, until the waters were dried up from off the earth.

Also he sent forth a dove from him, to see if the waters were abated from off the face of the ground;

But the dove found no rest for the sole of her foot, and she returned unto him into the ark, for the waters were on the face of the whole earth: then he put forth his hand, and took her, and pulled her into him into the ark.

And he stayed yet another seven days; and again he sent forth the dove out of the ark;

And the dove came in to him in the evening: and lo, in her mouth was an olive leaf pluckt off; so Noah knew that the waters were abated from off the earth. (Genesis, 8, vi-xi)

The three essential components of a bioassay are present in this example: a *stimulus* (the depth of water); a *biological test subject* (the dove); and a *response* (the plucking of an olive leaf).

Formal bioassays were developed to study the potency of insecticides during the early twentieth century at Rothamstead Station in England. Sir R. A. Fisher and other statisticians developed experimental designs and basic statistical procedures in collaboration with toxicologists and entomologists (Bliss, 1934a,b; Finney, 1952a; Gaddum, 1933).

Since that time, the role of bioassays has been expanded considerably. Prior to the 1970s, bioassays were used only to measure the toxicity of particular chemicals in, for example, medical, pharmacological, or agricultural studies (McKee and Wolf, 1963; Sprague, 1969). During the 1970s, the EPA began to use bioassay results for particular chemicals to establish the water-quality criteria that are published in the EPA green, blue, red, and gold books (e.g., United States Environmental Protection Agency, 1973, 1986). Since then, the application of bioassays has increased to include testing the toxicity of novel chemical compounds, licensing manufactured chemicals already in use (Federal Insecticide Fungicide Rodenticide Act), measuring toxicity at superfund sites (Resource Conservation and Recovery Act), and, of most importance for this paper, testing the toxicity of effluents.

EFFLUENT BIOASSAYS

The toxicity of effluents can be assessed by exposing test organisms to a series of dilutions of an effluent and measuring the organisms' responses. There are two basic classes of effluent bioassays. Acute bioassays measure survival of the target organisms. Chronic bioassays measure their growth, reproduction, or behavior. The toxic concentration of an effluent is defined as the lowest concentration of the effluent that causes a statistically significant effect on survival, growth, reproduction, or behavior, compared with a control.

Initial EPA-approved effluent bioassays were acute toxicity tests (Peltier, 1978). Beginning in 1985, EPA introduced guidelines for chronic toxicity tests. Chronic toxicity tests are important tools because they enable the detection of toxic effects that although sublethal may have important consequences for individuals exposed to the toxins. For example, a toxin could destroy an adult organism's ability to reproduce without killing the organism itself. Acute toxicity tests would fail to detect such effects. As the worst cases of environmental pollution have been detected and addressed, chronic tests have become increasingly important. The EPA has approved chronic toxicity tests that measure growth rates of algae populations, growth rates of larval fishes, and reproduction of small freshwater crustacean zooplankton (Horning and Weber, 1985). Because effluent bioassays have become the primary tool for ensuring the protection of ecosystems from toxic effluents, it is important to evaluate whether they achieve this goal.

REGULATORY APPEAL OF EFFLUENT BIOASSAYS

From an administrative viewpoint, regulations based on end-of-the-pipe measurements hold distinct advantages. These measurements (based on samples collected from pipes as they leave corporate or government facilities) require no information on the composition of the effluent or on when various constituents are added to the effluent. In addition, it is inherently easier to study the toxicity of effluents by working with the effluents themselves rather than with water from the receiving ecosystems after the addition of the effluents in question and any other inputs.

Acute bioassays have particular appeal because they are discrete, well-designed tests with well-established statistical procedures for characterizing dose-response effects (Bliss, 1934a,b; Finney, 1952a). If technicians follow required protocols carefully, they can perform acute tests accurately with minimal training. Chronic tests are more complicated. Because the performance of the test individuals in chronic tests is particularly sensitive to subtleties of culture conditions, food sources, and interactive effects of foods and toxins, technicians require more training and experience to perform chronic toxicity tests than acute toxicity tests. As a result, chronic tests are about three times as costly.

The National Pollution Discharge Elimination System (NPDES) administers the effluent bioassay program in individual states. The NPDES effluent toxicity standards are based on established state or federal water-quality criteria for compounds in effluents. Current recommended test species for freshwater effluents include the alga *Selenastrum capricornutum*, the crustacean zooplankton *Daphnia magna* and *Ceriodaphnia dubia*, and the fathead minnow, *Pimephales promelas*. Although other taxa can be used in effluent bioassays, most states require strict adherence to EPA protocols and quality assurance and quality control procedures.

Are the Results of Effluent Bioassays Misused?

Bioassays are effective tools for testing the toxicity of particular chemicals dissolved in water. They can also aid efforts to identify specific toxic compounds (Toxicity Identification Evaluations) or remove toxins from effluents (Toxicity Reduction Evaluations) (Mount and Anderson-Carnahan, 1988, 1989a,b). However, two important problems limit the use of bioassays as tools for protecting ecosystems from toxic effluents. First, the approved protocols present technical problems in application. Second and most important, effluent bioassay results do not enable the prediction of the effluent's impacts on other organisms or on ecosystem processes. In other words, acceptable toxicity in a bioassay is no guarantee that an effluent will not adversely affect the receiving habitat.

Technical Problems

The results of effluent bioassays are sensitive to multiple variables unrelated to the effluents themselves, including genetic variation in test organisms, chemical composition of the water source used to dilute the effluent, and foods used in experiments. Therefore, bioassay results may not be reproducible if these variables are not held constant. In such cases, EPA may not be able to confirm the results of tests performed by effluent generators or other laboratories. However, the EPA must be able to reproduce test results submitted by independent laboratories if they are to carry out their enforcement responsibilities. Consequently, EPA protocols specify the use of "artificial waters" (solutions that combine distilled water with various salts in an effort to mimic a pristine, natural, standardized water source), particular food sources (e.g., commercially available fish food), and even particular clones (genotypes) of the test species (Baird et al., 1989; Finney, 1952a). The detailed specification of testing protocols has forced many entities subject to EPA regulations to hire independent testing laboratories to perform their tests. The resulting costs have led to widespread dissatisfaction with effluent bioassay requirements.

Extrapolation from Test Results to Ecosystem Impacts

The second and more important shortcoming of effluent bioassays is the plethora of assumptions required to extrapolate test results to effects on receiving ecosystems. Neither a theoretical basis nor comprehensive test data validate the assumption that effluents that pass bioassays will not adversely affect receiving ecosystems. Such effluents may still harm ecosystems for either of two reasons: Test organisms may be less sensitive than other species to toxins in effluents, or the operational definition of acute or chronic toxicity may not account for certain adverse effects.

The species now used in EPA-approved bioassays were originally chosen based on the ease of culturing them and the availability of data on their sensitivities to various compounds. The same taxa were used to develop federal and state water-quality criteria. Even though these organisms were generally the most sensitive of the species evaluated, no data support the assumption that they are more sensitive than all other species in receiving ecosystems. In fact, most studies based on laboratory cultures use "weed" species—species that are easy to work with because they are relatively insensitive to changes in physical and chemical conditions. Consequently, it is likely that species evaluated as candidates for EPA-approved protocols had below-average sensitivity to changes in environmental conditions. An added complication exists when test protocols specify particular clones. Using clones increases the reproducibility of results but sacrifices information on the genetic variation within populations. Because clones can vary in sensitivity to toxins, it is unlikely that the clones used in bioassays are the most sensitive. With time, new, more-sensitive test species have been approved, but the selection process is generally driven by a combination of convenience considerations rather than a desire to identify the most sensitive species.

To compensate for the possibility that sensitive species are not protected, effluent bioassay protocols use "application factors" to calculate acceptable effluent concentrations for compounds. Application factors are essentially safety factors that reduce permitted effluent limits below those that show toxicity in bioassays (e.g., Peltier and Weber, 1985, p. 79). For example, if the LC_{50} (the concentration that kills 50 percent of test organisms) of an effluent is 1 part per million, a permit may require that the effluent concentration in the receiving ecosystem remain below 0.01 part per million. However, there is little biological or theoretical basis for choosing application factors. They may be either excessively or insufficiently protective (Forbes and Forbes, 1993).

Even if bioassays used the most-sensitive species as test organisms, several features of effluent bioassays would nevertheless complicate the use of test results as predictors of community- and ecosystem-level consequences of effluent releases. Most effluent bioassays measure changes in the survival, growth rate, reproduction, or behavior of individuals. However, these measures are insufficient to predict population dynamics in a taxon such as *Ceriodaphnia* because they do not measure density-dependent feedbacks. For example, feedback mechanisms may modify reproductive behavior at high population densities and low food concentrations, or high infant mortality may be offset by modifications in numbers of eggs produced or in the sizes of and nutrients present in individual eggs. Such effects are not incorporated into existing bioassay protocols.

Although it is difficult to make quantitative predictions of changes in population dynamics on the basis of toxicity to individuals, it is even more difficult to extrapolate further to community- and ecosystem-level consequences of toxicity. Suter et al. (1985, p. 400) describe the series of extrapolations involved in trying to predict community- and ecosystem-level consequences based on measures of an effluent's toxicity to individuals.

> The LC_{50} must be extrapolated from the test species to the species of interest, to life-cycle toxicity, to long-term toxicity in the field, to changes in population size due to direct toxic effects and, finally, to the combined direct and indirect toxic effects. Similarly, the emission rate must be converted into an effective environmental concentration in an imperfectly known hydrologic, chemical, physical, and biological system.

Ecology has not reached the point where such projections can be made with confidence. As a result, we are generally not able to make comprehensive quantitative predictions about the community- and ecosystem-level consequences of changes in, for example, the feeding rate of one member of the biotic community (Golley, 1994). In fact, many would argue that we are not even able to make confident extrapolations from bioassay data to consequences for field populations of the test organisms themselves.

A 1981 report from the National Research Council (NRC), *Testing for Effects of Chemicals on Ecosystems*, suggested that more-effective tests might be possible if they incorporated

> ... a significant number of species representing the degree of diversity found in the ecosystem, detailed observations on physiological and behavioral responses for individual species, [and] a time period similar to the duration of expected chemical exposure in the ecosystem. (p. 7)

However, such a "multispecies microcosm" approach has its own problems, both because the test conditions are oversimplified relative to real ecosystems and because the test conditions are more complex than those of single-species toxicity tests. Because microcosms are, of necessity, simplifications of actual ecosystems, they do not allow for all of the potential pathways of toxic effects that could occur in actual ecosystems. For example, few microcosm designs are large enough to include the largest organisms that occur in the natural ecosystems that the microcosms are intended to represent. Therefore, such tests can neither detect effects on those missing organisms, nor can they detect indirect effects that involve those missing organisms. However, because the tests involve more variables than do single-species bioassays, the mechanisms of any observed effects are more difficult to determine in multispecies microcosms than in single-species toxicity tests.

All of the above criticisms of effluent bioassays were identified by the authors of that same NRC review, which concluded that

Current laboratory tests examine only the responses of individuals, which are then averaged to give a mean response for the test species. With given constraints of limited finances and number of personnel, it is not possible to identify the most sensitive species or group of species. . . . The data are too limited in scope for extrapolations to be made from them to responses of other (even closely related) species. (pp. xi-xii)

After more than 15 years, we still do not know whether effluent bioassays sufficiently protect species in the field from direct toxic effects, and we do not have well-established methods for extrapolating from single-species toxicity measurements to community- and ecosystem-level effects of effluents. Although this is perhaps not surprising given the complexity of ecosystems and the number of variables involved, it must be recognized that this limitation in our understanding severely limits our ability to reliably extrapolate from the results of single-species bioassays to effects on receiving ecosystems.

SUMMARY

It is important to define the proper role of single-species bioassays. Single-species bioassays are suitable for developing water-quality criteria for particular chemicals, based on the assumption that these criteria protect individuals and that no synergistic effects occur with other chemicals in the environment. They are also useful as an initial screen to detect effluent toxicity. However, bioassays alone cannot ensure that effluents will not harm the ecosystems into which they are released.

REFERENCES

Baird, D. J., I. Barber, M. Bradley, P. Calow, and R. Soares. 1989. The *Daphnia* bioassay: A critique. Hydrobiologia 188/189:403–406.

Bliss, C. I. 1934a. The method of probits. Science 79:38–39.

Bliss, C. I. 1934b. The method of probits—A correction. Science 79:409–410.

Finney, D. J. 1952a. Statistical Method in Biological Assay. New York: Hafner Publishing.

Finney, D. J. 1952b. Probit Analysis, 2nd ed. Cambridge, England: Cambridge University Press.

Forbes, T. L., and V. E. Forbes. 1993. A critique of the use of distribution-based extrapolation models in ecotoxicology. Functional Ecology 7:249–254.

Gaddum, J. H. 1933. Reports on Biological Standards. III. Methods of Biological Assay Depending on Quantal Response. Special Report Series of the Medical Research Council, No. 183. London: Her Majesty's Stationery Office.

Golley, F. 1994. A History of the Ecosystem Concept in Ecology. New Haven, Conn.: Yale University Press.

Horning, W. B., and C. I. Weber, eds. 1985. Short-Term Methods for Estimating the Chronic Toxicity of Effluents and Receiving Waters to Freshwater Organisms. EPA/600/4-85/014. Environmental Monitoring and Support Laboratory. Cincinnati, Ohio: U.S. Environmental Protection Agency.

McKee, J. E., and H. W. Wolf. 1963. Water Quality Criteria, 2nd ed. Pub. No. 3A. Sacramento: California State Water Quality Control Board.

Mount, D. I., and L. Anderson-Carnahan. 1988. Methods for Aquatic Toxicity Identification Evaluations. Phase 1. Toxicity Characterization Procedures. EPA-600/3–88/034. Technical Report 02-88. Duluth, Minn.: National Effluent Toxicity Assessment Center.

Mount, D. I., and L. Anderson-Carnahan. 1989a. Methods for Aquatic Toxicity Identification Evaluations. Phase 2. Toxicity Identification Procedures. EPA-600/3-88/035. Technical Report 02-88. Duluth, Minn.: National Effluent Toxicity Assessment Center.

Mount, D. I., and L. Anderson-Carnahan. 1989b. Methods for Aquatic Toxicity Identification Evaluations. Phase 3. Toxicity Confirmation Procedures. EPA-600/3-88/036. Technical Report 04-88. Duluth, Minn.: National Effluent Toxicity Assessment Center.

National Research Council. 1981. Testing for Effects of Chemicals on Ecosystems. Washington, D.C.: National Academy Press.

Peltier, W. 1978. Methods for Measuring the Acute Toxicity of Effluents to Aquatic Organisms, 2nd ed. EPA/600/4-78/012. Environmental Monitoring and Support Laboratory. Cincinnati, Ohio: United States Environmental Protection Agency.

Peltier, W. H., and C. I. Weber, eds. 1985. Methods for Measuring the Acute Toxicity of Effluents to Freshwater and Marine Organisms, 3rd ed. EPA/600/4-85/013. Environmental Monitoring and Support Laboratory. Cincinnati, Ohio: U.S. Environmental Protection Agency.

Sprague, J. B. 1969. Measurement of pollutant toxicity to fish. I. Bioassay methods for acute toxicity. Water Research 3:793–821.

Suter II, G. W., L. W. Barnthouse, J. E. Breck, R. H. Gardner, and R. V. O'Neill. 1985. Extrapolating from the laboratory to the field: How uncertain are you? Pp. 400–413 in Aquatic Toxicology and Hazard Assessment: Seventh Symposium, R. D. Cardwell, R. Purdy, and R. C. Bahner, eds. ASTM STP 854. Philadelphia, Pa.: American Society for Testing and Materials.

United States Environmental Protection Agency. 1973. Water Quality Criteria 1972. EPA-R3–003. Washington, D.C.: U.S. Environmental Protection Agency.

United States Environmental Protection Agency, Office of Water Regulations and Standards. 1986. Quality Criteria for Water 1986. EPA 440/5-86-001. Washington, D.C.: U.S. Environmental Protection Agency.

Measures of Environmental Performance and
Ecosystem Condition. 1999. Pp. 199–216.
Washington, DC: National Academy Press.

Insights from Ambient Toxicity Testing

ARTHUR J. STEWART

INTRODUCTION

Ambient toxicity tests assess the toxicity of stream or river water by exposing organisms to the water and measuring their survival, growth, or reproduction. The performance of organisms in ambient toxicity tests can thus be used to directly assess the biological quality of waters that receive industrial or other effluents. This paper examines the types of insights that can be derived from ambient toxicity testing, based on lessons learned from several large-scale ambient toxicity testing programs established for streams that receive effluent from U.S. Department of Energy (DOE) facilities near Oak Ridge, Tenn.

In contrast to ambient toxicity tests that expose organisms to stream or river water, effluent toxicity tests (Goulden, this volume) expose organisms directly to effluent or diluted effluent. Regulations frequently require the use of effluent toxicity tests to document the biological quality of receiving waters. Standard methods approved by the U.S. Environmental Protection Agency (EPA) are available for both effluent and ambient toxicity tests (Kszos and Stewart, 1992; Weber et al., 1989).

Effluent and ambient toxicity tests use similar procedures but have different objectives. Both use "reagent grade" organisms as biodetectors, under standardized conditions, to provide a direct assessment of water quality. A subtle but important difference between effluent and ambient testing is this: In effluent testing, the key objective is usually to determine *how* toxic an effluent is, whereas in ambient testing, the main objective is usually to determine *whether* the water is toxic. A clear understanding of the differences between the two is necessary to design statistically rigorous, cost-effective, ambient toxicity testing programs.

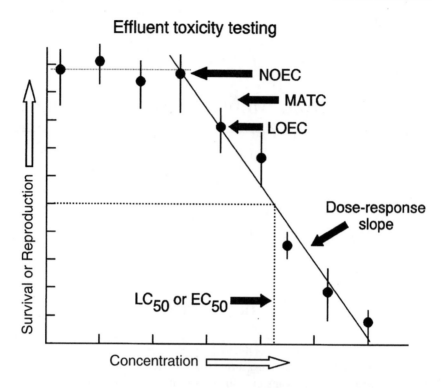

FIGURE 1   Generalized dose-response relationship suitable for estimating no-observed-effect concentration (NOEC), lowest-observed-effect concentration (LOEC), maximum-allowable toxicant concentration (MATC, equal to mean of NOEC and LOEC values), and LC_{50} and EC_{50} values (concentrations needed to kill 50 percent of the test organisms, or reduce the response variable by 50 percent, respectively).

In effluent testing, organisms are reared in various dilutions of effluent for a specified period of time with specified food, temperature, and light conditions. The ability of the organisms to survive, grow, and (in some cases) reproduce is measured and compared with the responses of organisms reared in a negative control (i.e., water known to be of good biological quality). The highest effluent concentration that causes no adverse effect is referred to as the no-observed-effect concentration (NOEC). The next-higher tested concentration, which shows the first statistically detectable effect of the effluent on the organisms, is referred to as the lowest-observed-effect concentration (LOEC). For regulatory purposes, effluent testing is used to establish a reliable estimate of an effluent's NOEC, LOEC, or LC_{50} (concentration of effluent that is lethal to half of the test organisms in a specified period of time) (Figure 1). The statistical procedures for estimating these concentrations are well defined (Figure 2). The NOECs can be compared with expected effluent concentrations in receiving streams to predict the likelihood of in-stream toxicity. Despite recent strong challenges to the concept of NOEC and LOEC on statistical grounds (see, for example, Kooijman, 1996), NOECs and LOECs are widely used and are likely to remain so for regulatory purposes in the United States for years to come.

The key difference between effluent toxicity data and ambient toxicity data may be best conceptualized in terms of signal-to-noise ratio. Compared with most receiving waters, most effluents have a strong toxicity "signal." On the other hand, the "noise level" for effluents tends to be lower than that of ambient waters. Thus, in general, the toxicity signal-to-noise ratio is higher for effluents than it is for ambient tests of receiving waters. This is important because it determines how the tests should be applied to maximize the information gained per dollar spent.

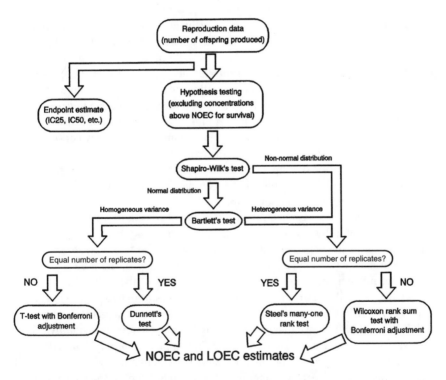

FIGURE 2 Statistical analysis flow path for reproduction data from *Ceriodaphnia* effluent toxicity tests (redrawn from Weber et al., 1989). NOTES: IC25 and IC50 refer to the concentrations of effluent or chemical that inhibits the measure of interest (such as growth or reproduction) by 25 or 50 percent, respectively. NOEC = no-observed-effect concentration. LOEC = lowest observed-effect concentration.

Waste treatment operators have a good understanding of their operations and know from experience and instrumentation feedback when treatment processes are operating correctly. In this situation, many water-quality conditions that can affect the outcome of a toxicity test (e.g., hardness, conductivity, suspended solids, pH, temperature) are relatively constant and predictable. In contrast, hardness, conductivity, and concentration of suspended solids can vary greatly in ambient waters with rainfall or snow-melt; pH can vary two standard units or more in response to season or even over daily cycles due to algal photosynthesis; and temperature can increase or decrease rapidly in response to weather conditions. Water-quality conditions in receiving streams can also change rapidly due to upstream spills or intermittent releases of batch-process effluents (e.g., cooling tower operations, which typically release a large volume of ion-rich waste water over a short period of time). In short, temporal variation in water quality is an important source of background noise that can complicate quantification of low levels of ambient toxicity.

Aquatic organisms are about as good at detecting toxicants in receiving waters as they are at detecting toxicants in effluents. However, the apparent or actual sensitivity of the organisms to some toxicants can be affected by other chemicals or water-quality factors. The sensitivity of test organisms to toxicants and their vulnerability to nontoxicant interferences are particularly important in ambient testing where the signal-to-noise ratio is low. Specific examples demonstrate this point. High but nontoxic concentrations of sodium can lower the toxicity of lithium to *Ceriodaphnia* (Stewart and Kszos, 1996). Thus, lithium at a concentration of 5 parts per million (ppm) in a sodium-rich waste water (e.g., 140 ppm sodium) might show no evidence of toxicity, whereas lithium is distinctly toxic at a concentration of 1 ppm in low-sodium (e.g., 5–10 ppm) ambient water. Calcium or other hardness-contributing materials can also lower the toxicity of nickel (Kszos et al., 1992) and other metals.

Physical variables also can affect the apparent sensitivity of test organisms. For example, naturally occurring particulate matter (algae, bacteria, and/or sediment) can lower the apparent sensitivity of organisms in two ways. First, some particulate matter (notably, algae, bacteria, and detritus) can be used as food by freshwater microcrustaceans. The nutritional benefits of the "extra food" can be important. For ambient toxicity tests of water samples from two sites in East Fork Poplar Creek (EFPC) (a stream that receives various waste waters from the DOE's Oak Ridge Y-12 Plant), we found that filtering the water to remove naturally occurring particulate matter significantly lowered *Ceriodaphnia* reproduction. The mean reduction in *Ceriodaphnia* reproduction caused by filtering the water was slightly larger at one site (9.7 percent) than it was at the other site (7.4 percent), but the effect of filtration was statistically significant at both sites ($p = 0.030$ for 15 tests at km 24.1 of EFPC; $p = 0.019$ for 21 tests at km 23.8 of EFPC, Student's T test). (Sites in East Fork Poplar Creek are identified by distance upstream from its confluence with Poplar Creek, a tributary of the Clinch River.) In contrast to many stream and river waters, most industrial effluents do not contain significant quantities of particulate matter because specific treatment operations such as polymer-enhanced flocculation or filtration remove solids from the water.

Particulate matter can also alter the bioavailability or biological activity of some contaminants. Chlorine (measured as total residual chlorine), for example, is very toxic to daphnids (Taylor, 1993) but is rapidly detoxified when it reacts with algae, detritus, or chemically labile dissolved organic matter. Organic and inorganic particles also are important sinks for relatively insoluble contaminants such as polychlorinated biphenyls, hydrophobic hydrocarbons, and various metals. In general, naturally occurring particulate matter lowers the concentrations of dissolved pollutants, thereby lowering the water's toxicity.

Monotonic response to an increase in the signal of interest is an important consideration in any detector system. The dose-response concept in toxicity testing embodies this consideration and has been very influential in the development of effluent toxicity tests. Dose-response patterns, where organism responses are a function of toxin concentration, are fundamental to effluent and pure-chemical toxicity testing. It is through adherence to an expected dose-response relationship that effluent toxicity testing gains predictive value. Therefore, much effort in the development of toxicity tests has gone into the selection of test procedures that generate smooth dose-response curves. Procedures for establishing regulatory limits on effluent toxicity are based on the premise that a monotonic dose-response relationship can be determined (Figure 1). This premise dominates every aspect of effluent toxicity testing: Adherence to a linear dose-response pattern allows extraction of the toxicity signal. The statistical procedures for estimating toxicity of effluents that yield smooth dose-response curves are clearly outlined in EPA manuals (Kooijman, 1996; Weber et al., 1989) (Figure 1).

A weak toxicity signal and relatively high background noise are typical of ambient toxicity test conditions. Low signal-to-noise ratios prevent effective quantification of ambient toxicity using the statistical framework of the dose-response model that works so well for effluent toxicity tests. The difference in appropriate statistical procedures for analysis of test results is the crucial distinction between the analytical procedures for ambient and effluent toxicity tests. This difference is central to the formulation of a cost-effective strategy for ambient toxicity testing.

APPLICATIONS

Ambient toxicity tests using *Ceriodaphnia dubia* (a freshwater microcrustacean) and *Pimephales promelas* (fathead minnow) larvae have been used to support biological monitoring programs for 12 receiving streams at DOE facilities in Oak Ridge, Tenn. and Paducah, Ky. (Stewart and Loar, 1994). The tests used EPA-approved procedures for estimating chronic toxicity (Mount and Norberg, 1984; Norberg and Mount, 1985; Weber et al., 1989), specifically, rearing replicate groups of fathead minnow larvae, or individual *Ceriodaphnia*, in full-strength (i.e., nondiluted) samples of water from the site(s) being evaluated. During the 7-day tests, fresh samples of water were collected daily from each site being evaluated. The water was then warmed to the test temperature (25°C) and used to replace the previous day's water in the test chambers. This procedure is referred to as static renewal of test cultures. In almost every case, we evaluated the sites for ambient toxicity by testing both species concurrently. The ambient toxicity tests include negative controls (i.e., tests with mineral water, diluted to an acceptable ionic strength with distilled water) and water samples from reference sites, located upstream of known point- or area-source inputs of pollutants. We estimated toxicity using survival and growth of minnow larvae and survival and reproduction of *Ceriodaphnia*. We also measured the pH, conductivity, alkalinity, hardness, and total residual chlorine of all freshly collected water samples.

Ambient toxicity tests were run on water from as many as 10 sites per stream, but in some cases one site was sufficient for effective monitoring. At one of the DOE facilities near Oak Ridge (the Oak Ridge National Laboratory), 15 sites on 5 receiving streams have been tested 42 times (concurrently with both species) since 1986. A total of 630 site and test-period combinations were represented by the sampling and testing strategy used for the five receiving streams (i.e., 15 sites in each of 42 test periods). For each stream or suite of streams that is monitored for ambient toxicity, the primary unit of statistical analysis is the mean response for each site-date combination. The response parameters include *Ceriodaphnia* survival (percentage, based on 10 animals per site-date combination) and reproduction (number of offspring per female, for females that survive all 7 days), and fathead minnow survival (percentage, based on 4 replicates, each containing 10 fish) and growth (mean milligrams dry mass per surviving fish, per replicate, corrected for initial weight). Survival data for the minnows are generally arc-sine square-root transformed before analysis; growth data for the minnow larvae can be corrected for growth of larvae in the controls or reference sites, depending on the objective of the analysis. We do not normally transform *Ceriodaphnia* survival values because each test generates only a single value (e.g., 100 percent, 90 percent, 80 percent) derived from 10 individual animals, each of which constitutes a replicate.

We have explored various methods for analyzing the results of ambient toxicity tests. The case-study examples below summarize these methods, the key findings revealed through their use, and their major advantages and disadvantages. Most of these examples are derived from studies published elsewhere (e.g., Kszos et al., 1992; Stewart, 1996; Stewart et al. ,1990, 1996). In general, the results of ambient tests are used either to reveal differences among sites (e.g., a longitudinal pattern within a stream or differences among streams) or to demonstrate the occurrence of water-quality changes over time.

Analysis of Variance Methods

Analysis of variance (ANOVA) can be conducted using site, test period, and the interaction between site and test period as explanatory factors for *Ceriodaphnia* survival or reproduction, or fathead minnow survival or growth. ANOVA (SAS-GLM, available for use on personal computers; SAS Institute, 1985) provides an estimate of the amount of variation in survival, reproduction, or growth that is explained (R^2) by the three factors together. Duncan's multiple-range test (a SAS-GLM option) or other multiple-comparison tests can be used to identify sites or test periods where the response factor is low. When Duncan's multiple-range test is used to identify differences among sites, sites are sorted according to mean responses. Sites that have an unusually low mean value for any of the four response parameters can be considered as suspect for toxicity. If the study involves a linear array of sites below a discharge, and the effects attributable to date and the interaction of site and test period are small, the procedure could permit the investigator to identify a "no-observed-effect site," analogous to the NOEC of effluent toxicity tests.

If data for a sufficiently large number of test periods are available, and one or two of the test periods have unusually low mean values for a response parameter, the data set can be pruned by eliminating data from the suspect test period(s). This procedure may be justified if the response parameter in question (e.g., fathead minnow growth) is unusually low in water from all sources, including references sites, and the control. The elimination of data for test periods that have suspiciously low values for the response factors should increase R^2 for the full model (site, test period, and the interaction between site and test period) and lower the significance of test period. An analysis reported in Boston et al. (1994) showed an increase in the amount of explained variance in *Ceriodaphnia* survival and reproduction by eliminating suspect dates from the data set. In contrast, neither the results for minnow survival or growth did not benefit much from data pruning. Pruning should be used only when it is thoroughly justified. In such cases, the justification should be explained, and the consequences of the act of pruning should be considered carefully. The objective of pruning is not to increase the R^2 of a linear model but to reveal temporal or spatial patterns in biological quality of the water that may otherwise be obscured by excessive variance due to test dates where growth, survival, or reproduction of test organisms was low in control or reference conditions.

When using toxicity test methods to assess ambient water quality, a bioassay should simultaneously meet two key objectives: It should discriminate readily among sites, and it should exhibit little variation from test period to test period, when applied to noncontaminated control water or to water from a noncontaminated reference site. An ANOVA-based analysis of results of 285 site and test-period combinations was used to determine which test organism—*Ceriodaphnia* or fathead minnow larvae—best fulfilled these objectives (Boston et al., 1994). That analysis considered rank values of sites within test periods and of test periods within sites for all four measures of toxicity (minnow survival and growth, and *Ceriodaphnia* survival and reproduction). The suitability of these two types of tests for ambient applications was evaluated in a two-step process. First, for each of the four measures of toxicity (dependent variables), a site specificity over time term was computed by dividing the proportion of variation in the dependent variable explained using site as the explanatory factor (i.e., the R^2 for site) by the proportion of variation explained by using test period as the explanatory factor (i.e., the R^2 for test period). The relative utility of each test organism was then computed by summing its two "site specificity over time" terms (one term for each measure of toxicity). This computation showed that the *Ceriodaphnia* test was about 3.8 times more specific than the fathead minnow test. One significant conclusion from the study by Boston et al. (1994) was that, for ambient water-quality assessments, greater testing frequency with *Ceriodaphnia* might be more effective than less frequent testing with both species.

The examples described above use ANOVA with site and test period as explanatory factors. This approach allows one to determine if site or test period has a statistically significant effect on fathead minnow larvae survival or growth, or on *Ceriodaphnia* survival or reproduction. These methods cannot be used to infer that toxicants cause a response, even if one site differs greatly from the others with respect to survival, reproduction, or growth. Other data must be considered to ascertain the cause of the observed response. Using two-way ANOVA, responses of organisms in the controls can be used qualitatively, to support the idea of pruning results from a particular test date from the data set, or quantitatively, as though controls were merely an additional site. When used in the latter fashion, some of the sites frequently appear significantly better than controls and some sites worse than the controls. In effluent testing, controls are essential; in ambient testing, controls are useful, but reference sites are critical.

Contingency-Table Analysis Methods

We have used contingency-table methods to establish lower-bound values for "passing" an ambient toxicity test of *Ceriodaphnia* survival. The procedure is simple and practical in concept and its computation is similar to Fisher's Exact Test, which EPA recommends for assessing *Ceriodaphnia* survival in effluent toxicity tests. The main drawback of contingency-based methods is that generating strong conclusions requires data from a large number of ambient tests at one or more reference sites.

Basically, the contingency-table method involves categorizing and tabulating *Ceriodaphnia* test results to reveal the distribution of survival values for ambient tests of water from several reference sites pooled through time. The distributions of the test outcomes can be used in two ways. First, they can be used as

TABLE 1   Distribution of Survival Values for *Ceriodaphnia*

	Ceriodaphnia Survival (percent)						Total Number of Tests
Water Source	100	90	80	70	60	≤50	
Diluted mineral water	30	21	9	2	0	0	62
First Creek 0.9 km	23	12	2	2	0	0	39
Fifth Creek 1.1 km	22	14	2	3	0	0	41
White Oak Creek 6.8 km	28	8	3	1	1	0	41
Reference sites combined	73	34	7	6	1	0	121

NOTE: Data from 7-day tests conducted using diluted mineral water and water samples from noncontaminated reference sites in three streams near the Oak Ridge National Laboratory. The last row of the table shows pooled results for the three reference sites.

a reference for identifying suspiciously low survival or mean reproduction values for tests of nonreference sites. Second, the distribution of survival values for reference-site tests can be compared formally with the distribution of survival values in control tests through application of an appropriate test (e.g., Chi-square).

Table 1 gives an example of the distribution of *Ceriodaphnia* survival values for controls and ambient tests of reference sites in three streams near the Oak Ridge National Laboratory. Inspection of this table shows that the distribution of *Ceriodaphnia* survival values in reference-site tests is very similar to their distribution in control water (diluted mineral water). Thus, the probability that a reference-site ambient-water test would yield a *Ceriodaphnia* survival value that is equal to or lower than 60 percent can be estimated as $100 \times (1 \text{ case} \div 121 \text{ cases})$ (see Table 1), or 0.008. Accordingly, if *Ceriodaphnia* survival is 50 percent in a 7-day test of water collected from a receiving stream, the low survival value is unlikely to be due to chance alone.

A contingency-table analysis method also could be used to establish a lower pass-or-fail criterion for *Ceriodaphnia* reproduction or fathead minnow survival or growth. Using data for reference sites in the three streams near Oak Ridge National Laboratory, we found that *Ceriodaphnia* mean reproduction values were less than or equal to 10 offspring per surviving female in only 6 of 121 tests (Table 2), or about 5 percent of the cases. Thus, one could use 10 offspring per surviving female as the lower-bound criterion for passing a *Ceriodaphnia* reproduction ambient toxicity test. However, within a given test, each surviving daphnid serves as a replicate and yields a value for reproduction. Replicate values are also available for fathead minnow survival and growth. The information from replicates permits the use of other, more powerful methods of analysis, such as ANOVA.

TABLE 2   Distribution of *Ceriodaphnia* Reproduction Values

Ceriodaphnia Reproduction

Water Source	≥30	≥25–30	≥20–25	≥15–20	≥10–15	<10	Total Number of Tests
Diluted mineral water	8	13	23	9	5	4	62
First Creek 0.9 km	6	6	13	9	3	2	39
Fifth Creek 1.1 km	4	6	12	11	6	2	41
White Oak Creek 6.8 km	3	10	10	14	2	2	41
Reference sites combined	13	22	35	34	11	6	121

NOTE: Values represent mean number of offspring per surviving female in 7-day tests of water from various sources. Diluted mineral water tests are used as negative controls. First Creek 0.9 km, Fifth Creek 1.1 km, and White Oak Creek 6.8 km are noncontaminated reference sites in streams near the Oak Ridge National Laboratory. The last row of the table shows pooled results for the three reference sites.

Assessment of Concordance Patterns

The ANOVA and contingency-table methods described above consider responses of a particular species (e.g., *Ceriodaphnia dubia* or *Pimephales promelas*) separately, in relation to site. Ambient toxicity data may also answer the question, Are site-to-site differences in responses for one species similar to the site-to-site differences in responses for a second species? A similar spatial response pattern for two or more species strengthens the argument that biological water quality is site specific. It is intuitively clear that a strongly polluted site should adversely affect various species and that various species should do "better" in water that lacks pollutants. For 180 site and test-period combinations (15 sites, 12 test periods) of five receiving streams at Oak Ridge National Laboratory, we found that the correlation between *Ceriodaphnia* survival and fathead minnow survival was positive and significant ($p < 0.0001$); *Ceriodaphnia* reproduction and fathead minnow growth were also correlated, but less strongly ($p = 0.026$) (Stewart et al., 1990). This finding supports the idea that testing more than one species has some value and provides evidence for the notion that biologically significant differences in water quality may be revealed through assessment of concordance.

One simple method for using concordance ranks sites according to responses for each species separately, then tabulates the number of cases in which each site is best or worst for each species, for either species, or for both species together. Table 3 shows an example of this approach, using *Ceriodaphnia* and fathead minnow toxicity test results for eight test periods and six sites in East Fork Poplar Creek. In this example, no site stood out as being consistently better or worse in terms of fathead minnow growth or *Ceriodaphnia* reproduction. This analysis showed no detectable longitudinal pattern to water quality in the stream, based on either fathead minnow growth or *Ceriodaphnia* reproduction in 7-day tests (Boston et al., 1993). Water from km 13.8 of East Fork Poplar Creek, however, appeared to be consistently better than water from the other sites: It was never the worst for either species and was the best for one or the other of the two species in six of the eight test periods.

Multivariate Analyses

The ANOVA, contingency-table, and concordance-pattern methods can be used to reveal biologically based water-quality differences among sites. A more powerful and predictive framework for the analysis of ambient toxicity test outcomes can be established by linking responses of the test organisms specifically to chemical measurements of water quality. Various statistical methods are available for this purpose. Examples of two such methods—principal components analysis followed by multiple regression analysis, and logistic regression—are summarized below.

Principal components analysis (PCA) and multiple regression analysis were used to inspect relationships between ambient toxicity test outcomes and chemical variables for 180 site and test-period combinations (15 sites each tested 12 times) in receiving streams near Oak Ridge National Laboratory (Stewart et al., 1990). *Ceriodaphnia* and fathead minnow larvae were tested concurrently in each test period. Chemical water-quality parameters measured for each site-date

TABLE 3   Results of Ambient Toxicity Tests of Water from Six Sites on East Fork Poplar Creek

	Site					
Water Quality	22.8	21.9	20.5	18.2	13.8	10.9
Best for minnow growth	2	1	3	1	2	1
Best for *Ceriodaphnia*	1	1	0	1	4	1
Best for both species[a]	0	1	0	0	2	0
Best for either species	3	2	3	2	6	2
Worst for minnow growth	1	3	2	0	0	2
Worst for *Ceriodaphnia*	2	1	1	3	0	2
Worst for both species[a]	0	1	0	0	0	1
Worst for either species	3	3	3	3	0	3

[a]In all cases *Ceriodaphnia dubia* and *Pimephales promelas* were tested concurrently.

NOTE: Numerals specifying sites refer to distances (km) upstream from the confluence of East Fork Poplar Creek with the Clinch River.

combination included pH, alkalinity, conductivity, hardness, and total residual chlorine (TRC). First, 7-day means for each water-quality factor were computed. PCA then was used to identify two orthogonal water-quality axes (axis I, associated primarily with hardness, conductivity, and pH; and axis II, strongly associated with TRC). The two axes accounted for 60.5 and 17.6 percent, respectively, of the total variance in the chemical data. Multiple regression analysis was then used to test relationships between the results of the ambient toxicity tests and the two principal component factors. This analysis showed that the fathead minnow survival and growth did not correspond well to any combination of the measured chemical variables and that the *Ceriodaphnia* test results related strongly to axes I and II. Mean survival of *Ceriodaphnia* was related strongly to axis II ($p <$ 0.001) and secondarily to axis I ($p = 0.101$), whereas mean reproduction of *Ceriodaphnia* had strong relationships to both axes (axis I, $p = 0.011$; axis II, $p = 0.019$). The results of the PCA–multiple regression analyses suggested that TRC was a biologically significant contaminant whose presence strongly influenced *Ceriodaphnia* test outcomes.

We were able to draw two more conclusions from the study using other supporting analyses. First, for ambient assessments of water quality in these streams, *Ceriodaphnia* tests detected toxic conditions better than fathead minnow tests. This conclusion was supported by examination of R^2 changes in ANOVAs of the *Ceriodaphnia* and fathead minnow tests in response to data pruning by date, as described above. Second, we were able to show that ambient toxicity dynamics in the Oak Ridge National Laboratory streams were dominated by episodic events that sometimes caused acutely toxic conditions at "poor quality" sites. Together, the three conclusions focused subsequent remediation activities and shaped the strategy for more cost-effective monitoring at the Oak Ridge National Laboratory. We began frequent testing to assess episodic events, but using *Ceriodaphnia* only for reasons of sensitivity and cost; we documented long-term improvements in water quality by monitoring biological and chemical conditions at the poor quality sites; to reduce costs, we halted testing at nonreference sites that have shown no evidence of toxicity; and we continue to conduct special studies and use diagnostic testing to better understand the fate and ecological effects of low concentrations of TRC.

Logistic regression analysis was used to relate chemical conditions to *Ceriodaphnia* mortality patterns in water samples from East Fork Poplar Creek. When using 7-day static-renewal toxicity test methods to assess ambient water quality, the water in the test chambers is replaced daily with freshly collected water. This procedure generates both an interesting challenge and a strong potential bias. The challenge is this: How should one best relate a time-varying exposure regime (e.g., daily changes in conductivity, pH, TRC) to a single, biologically integrated measure of "response" (e.g., *Ceriodaphnia* reproduction, expressed as a 7-day mean)? The potential bias also relates to the problem of time. The physicochemical characteristics of a sample of stream water may not be representative of in situ physicochemical conditions because some parameters (such as pH level) can vary naturally over daily cycles, and others (such as conductivity) may change strongly in response to waste-water discharges. These two issues were explored by using *Ceriodaphnia* tests to evaluate water-quality conditions in upper East Fork Poplar Creek, where TRC was suspected of causing or contributing to fish kills. Logistic regression was used to relate TRC data to toxicity test outcomes (Stewart et al., 1996).

We first analyzed the chemical data (daily measurements of pH, conductivity, alkalinity, hardness, and TRC) for 169 site and test-period combinations (4 sites were tested over a 50-month period). For each water-quality factor, we computed a 7-day mean and an estimate of daily variability, referred to as semirange. Semirange was defined as a parameter's 7-day maximum (transformed) value minus the 7-day mean. For toxicity assessments, one advantage of semirange is that it quantifies excursions above the mean but ignores excursions below the mean. (Toxicologically, pollutant concentrations above the mean are likely to be more significant than those below.) We then used stepwise logistic regression to explore relationships between the 7-day mean and 7-day semirange values for the water-quality factors and *Ceriodaphnia* mortality. Both the proportion of animals dying in each test and the pass-or-fail outcomes (using 60 percent survival as the pass-or-fail criterion [see Table 1]) were assessed.

The results of these analyses showed that 7-day mean TRC concentration and TRC semirange both strongly affected *Ceriodaphnia* mortality ($p < 0.0001$ for each factor). With these two factors included, the logistic regression model correctly predicted the outcome (mortality or survival, expressed as a proportion of the animals tested in each test) in 89.3 percent of the cases. The model's false positive rate (when the model predicted mortality, but no mortality occurred) was 20 percent, and the model's false negative rate (no mortality was predicted by the model but the animal died) was 7 percent. Distilling a test's outcome to a pass-or-fail status using the criterion of 60 percent survival was a satisfactory simplification: Both TRC mean and semirange values were significant as explanatory factors ($p < 0.0001$ in each case, with 91.7 percent of the cases being predicted correctly by the model), and the model's false positive rate and false negative rates were low (15.2 and 5.7 percent, respectively).

Figure 3 is a schematic showing the generalized flow for *Ceriodaphnia* toxicity test data used in the statistical analysis methods described in this paper. Various data-checking steps cited for use in the effluent data flow path (e.g., inspection of variance for homogeneity [Figure 2]) are also appropriate when analyzing ambient toxicity test data, but these are not shown in Figure 3 for convenience.

Diagnostic Testing and Ambient Toxicity Monitoring

The logistic regression study described above also demonstrated that diagnostic or "experimental" toxicity testing should be integrated into any ambient

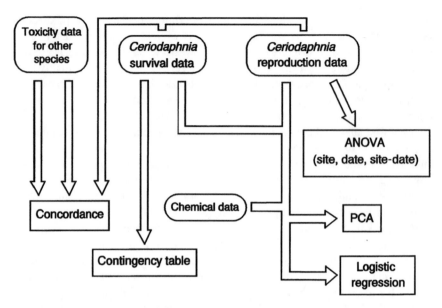

FIGURE 3   Generalized statistical analysis flow path for survival and reproduction data from *Ceriodaphnia* toxicity tests used for ambient water-quality monitoring (details provided in text).

toxicity monitoring program. Concurrent with the routine ambient toxicity monitoring tests for upper East Fork Poplar Creek, we conducted diagnostic toxicity tests to demonstrate that TRC (or related oxidants) accounted for the observed toxicity. In diagnostic testing, a specific treatment is imposed to alter water quality, and organisms' responses are compared statistically with those of the organisms in nontreated water to demonstrate causality. In the logistic regression study, diagnostic testing consisted of comparing responses of *Ceriodaphnia* in samples of dechlorinated water to nontreated water (daily renewal of water in both cases), with dechlorination using small quantities of sodium thiosulfate. Other treatments used in diagnostic ambient testing include adding metal-complexation agents (e.g., ethylenediaminetetraacetic acid); filtering to remove particulate matter; exposing the water to strong ultraviolet light to alter photosensitive chemicals or kill bacteria; aerating the water; adjusting the pH; and passing the water through a column of activated carbon. The results of the side-by-side tests of treated and nontreated water samples can be analyzed easily and effectively by ANOVA, with separate test periods serving as replicates. Examples of effective application of diagnostic experiments conducted to support ambient water-quality assessments are provided in studies by Kszos and Stewart (1992), Kszos et al. (1992), Nimmo et al. (1990), and Stewart et al. (1996).

Artifacts in Ambient Testing

Factors other than toxicants can affect fathead minnow survival and growth and *Ceriodaphnia* survival and reproduction in ambient waters. Growth of minnow larvae in laboratory tests, for example, is affected by concentrations of common salts, and survival of the larvae in ambient waters from relatively pristine streams can be low and variable due to the presence of pathogenic microorganisms (Kszos et al., 1997). *Ceriodaphnia* reproduction is commonly greater in ambient water than in diluted mineral-water controls, due to the nutritional benefits they derive from consuming naturally occurring particulate matter, but some naturally occurring algae can be toxic (Reinikainen et al., 1994). These situations make it inadvisable to compare the results of ambient tests only with diluted mineral water controls to determine if an ambient site is toxic or nontoxic. Comparison with an appropriate suite of reference sites is critical to derive the correct answer for the correct reasons. In ambient toxicity testing, and in biological monitoring generally, one must be constantly alert to the difference between biological importance and statistical significance (Cairns and Smith, 1994; Yoccoz, 1991).

PATH FORWARD

New and potentially useful ambient assessment procedures are being developed at a rapid pace; innovations in biological monitoring occur more slowly; slower still is the rate at which field-validated bioassessment methodology is being incorporated and used in a regulatory framework (see, for example, Hart, 1994). Examples of rapid progress in bioassay development can be found in both the water- and the soil-assessment arenas. A 3-day laboratory test that uses snail feeding rate to evaluate water quality appears to be about as sensitive as a 7-day *Ceriodaphnia* test, at least for some kinds of contaminants (R. L. Hinzman, Environmental Sciences Division, Oak Ridge National Laboratory, unpublished data). Procedures for estimating the toxicity of sediments with laboratory tests using invertebrates are nearing readiness for regulatory use (American Society for Testing and Materials, 1991). Methods for laboratory tests designed to estimate the toxicity of soils are being revised, calibrated, and field validated (L. F. Wicker, Environmental Sciences Division, Oak Ridge National Laboratory, unpublished data).

The increasing use of ecological risk assessment methodology for regulatory purposes drives the need not only for faster and more cost-effective laboratory tests, but also for data that accurately reflect exposure regimes and reveal ecological effects in the field. In situ test procedures using caged or noncaged organisms are in various stages of development and validation for terrestrial (Callahan et al., 1991; Menzie et al., 1992) and aquatic environments (Napolitano et al., 1993). Aquatic (Graney et al., 1994) and terrestrial (Gunderson et al., 1997; Parmalee et al., 1993) mesocosm studies are key to the development of in situ test methods that ultimately will be required for effective use of ecological risk-assessment methodology. Despite their limitations, simple laboratory tests, such as the *Ceriodaphnia dubia* and *Pimephales promelas* tests described in this paper, are likely to be relied on more and more. This is because the need for data that can be used for ecological risk assessments grows much faster than the rate at which regulatory agencies approve new methods for assessing the environment.

CONCLUSIONS

Standardized tests designed to estimate the toxicity of effluents to aquatic biota can, with minimal modification, also be used to assess ambient water-quality conditions in receiving streams. However, ambient tests should not be analyzed statistically in the same way as effluent tests. Site and test-period combinations, rather than effluent concentrations, serve as the principal unit of assessment for ambient toxicity test results. In addition, the results of ambient tests are in many cases more appropriately compared with the results of reference-site tests than with negative controls, which are commonly included with effluent tests. These considerations shape the strategy for cost-effective use of ambient toxicity testing. The value of ambient toxicity testing increases if the tests are used to support a broader-based, long-term biological monitoring program; conducted frequently with one sensitive species, rather than more often with two or more species; and accompanied by a diagnostic ("experimental") toxicity testing program. Data pruning by date can be used to help identify sites where water quality is suspect, and a representative suite of reference sites should be included in every ambient testing program to help place suspect sites into appropriate perspective. Specific linkages between ambient toxicity test results and chemical conditions at the test site are extremely desirable and can be revealed using methods such as PCA or logistic regression. The long-term prognosis is that in situ testing will replace the ambient toxicity testing procedures now in use. However, requirements for data that can be used in ecological risk assessments are likely to grow much faster than the rate of approval for in situ test methods for regulatory purposes. Thus, the next decade is likely to bring a marked increase in ambient testing with EPA-approved static-renewal laboratory procedures using organisms such as *Ceriodaphnia* and fathead minnow larvae.

ACKNOWLEDGMENTS

This paper was improved through reviews and comments provided by T. L. Ashwood and T. L. Phipps and was made possible by technical contributions from members of the Biomonitoring Group, including L. A. Kszos, T. L. Phipps, L. F. Wicker, P. W. Braden, G. W. Morris, B. K. Beane, L. S. Ewald, J. R. Sumner, K. J. McAfee, and W. S. Session. Oak Ridge National Laboratory is managed for the U.S. Department of Energy by Lockheed Martin Energy Research Corp. under contract DE-AC05–96OR22464. The Oak Ridge Y-12 Plant is managed for the U.S. Department of Energy by Lockheed Martin Energy Systems, Inc., under contract DE-AC05–84OR21400.

REFERENCES

American Society for Testing and Materials. 1991. Standard Guide for Conducting Sediment Toxicity Tests with Freshwater Invertebrates. Philadelphia, Pa.: American Society for Testing and Materials.

Boston, H. L., W. R. Hill, and A. J. Stewart. 1993. Toxicity monitoring. Pp. 37–108 in Second Report on the Oak Ridge Y-12 Plant Biological Monitoring and Abatement Program for East Fork Poplar Creek, R. L. Hinzman, ed. Y/TS-888. Oak Ridge, Tenn.: Environmental Sciences Division, Oak Ridge National Laboratory.

Boston, H. L., W. R. Hill, L. A. Kszos, C. M. Pettway, and A. J. Stewart. 1994. Toxicity monitoring. Pp. 27–63 in Fourth Report on the Oak Ridge National Laboratory Biological Monitoring and Abatement Program for White Oak Creek and the Clinch River, J. M. Loar, ed. ORNL/TM-11544. Oak Ridge, Tenn.: Environmental Sciences Division, Oak Ridge National Laboratory.

Cairns, J., Jr., and E. P. Smith. 1994. The statistical validity of biomonitoring data. Pp 49–68 in Biological Monitoring of Aquatic Systems, S. L. Loeb and A. Spacie, eds. Boca Raton, Fla.: Lewis Publishers.

Callahan, C. A., C. A. Menzie, D. E. Burmaster, D. C. Wilborn, and T. Ernst. 1991. On-site methods for assessing chemical impact on the soil environment using earthworms: A case study at the Baird & McGuire Superfund Site, Holbrook, Massachusetts. Environmental Toxicology and Chemistry 10:817–826.

Graney, R. L., D. H. Kennedy, and J. H. Rodgers, eds. 1994. Aquatic Mesocosm Studies in Ecological Risk Assessment. Boca Raton, Fla.: Lewis Publishers.

Gunderson, C. A., J. M. Kostuk, M. H. Gibbs, G. E. Napolitano, L. F. Wicker, J. E. Richmond, and A. J. Stewart. 1997. Multispecies toxicity assessment of compost produced in bioremediation of an explosives-contaminated sediment. Environmental Toxicology and Chemistry 16(12):2529–2537.

Hart, D. D. 1994. Building a stronger partnership between ecological research and biological monitoring. Journal of the North American Benthological Society 13:110–116.

Kooijman, S. A. L. 1996. An alternative for NOEC exists, but the standard model has to be abandoned first. Oikos 75:310–316.

Kszos, L. A., and A. J. Stewart. 1992. Artifacts in ambient toxicity testing. Paper AC92–020–005 in Proceedings of the Water Environment Federation, 65th Annual Conference and Exposition, New Orleans, La., September 20–24.

Kszos, L. A., A. J. Stewart, and P. A. Taylor. 1992. An evaluation of nickel toxicity to Ceriodaphnia dubia and Daphnia magna in a contaminated stream and in laboratory tests. Environmental Toxicology and Chemistry 11:1001–1012.

Kszos, L. A., A. J. Stewart, and J. R. Sumner. 1997. Evidence that variability in ambient fathead minnow short-term chronic tests is due to pathogenic infection. Environmental Toxicology and Chemistry 16:351–356.

Menzie, C. A., D. E. Burmaster, J. S. Freshman, and C. A. Callahan. 1992. Assessment of methods for estimating ecological risk in the terrestrial component: A case study at the Baird & McGuire Superfund Site in Holbrook, Massachusetts. Environmental Toxicology and Chemistry 11:245–260.

Mount, D. I., and T. J. Norberg. 1984. A seven-day life cycle cladoceran toxicity test. Environmental Toxicology and Chemistry 3:425–434.

Napolitano, G. E., W. R. Hill, J. B. Guckert, A. J. Stewart, S. C. Nold, and D. C. White. 1993. Changes in periphyton fatty acid composition in chlorine-polluted streams. Journal of the North American Benthological Society 13:237–249.

Nimmo, D. R., M. H. Dodson, P. H. Davies, J. C. Greene, and M. A. Kerr. 1990. Three studies using *Ceriodaphnia* to detect nonpoint sources of metals from mine drainage. Journal of the Water Pollution Control Federation 62:7–15.

Norberg, T. J., and D. I. Mount. 1985. A new subchronic fathead minnow (*Pimephales promelas*) toxicity test. Environmental Toxicology and Chemistry 4:711–718.

Parmalee, R. W., R. S. Wentsel, C. T. Phillips, M. Simini, and R. T. Checkai. 1993. Soil microcosm for testing the effects of chemical pollutants on soil fauna communities and trophic structure. Environmental Toxicology and Chemistry 12:1477–1486.

Reinikainen, M., M. Ketol, and M. Walls. 1994. Effects of the concentrations of toxic *Microcystis aeruginosa* and an alternative food on the survival of *Daphnia pulex*. Limnology and Oceanography 39:424–432.

SAS Institute. 1985. SAS User's Guide: Statistics. Version 5. Cary, N.C.: SAS Institute.

Stewart, A. J. 1996. Ambient bioassays for assessing water-quality conditions in receiving streams. Ecotoxicology 5:377–393.

Stewart, A. J., and L. A. Kszos. 1996. Caution on using lithium (Li+) as a conservative tracer in hydrological studies. Limnology and Oceanography 41:190–191.

Stewart, A. J., and J. M. Loar. 1994. Spatial and temporal variation in biomonitoring data. Pp. 91–124 in Biological Monitoring of Aquatic Systems, S. Loeb and A. Spacie, eds. Boca Raton, Fla.: Lewis Publishers.

Stewart, A. J., W. R. Hill, K. D. Ham, S. W. Christensen, and J. J. Beauchamp. 1996. Chlorine dynamics and ambient toxicity in receiving streams. Ecological Applications 6:458–471.

Stewart, A. J., L. A. Kszos, B. C. Harvey, L. F. Wicker, G. J. Haynes, and R. D. Bailey. 1990. Ambient toxicity dynamics: Assessments using *Ceriodaphnia dubia* and fathead minnow (*Pimephales promelas*) larvae in short-term tests. Environmental Toxicology and Chemistry 9:367–379.

Taylor, P. A. 1993. An evaluation of the toxicity of various forms of chlorine to *Ceriodaphnia dubia*. Environmental Toxicology and Chemistry 12:925–930.

Weber, C. I., W. H. Peltier, T. J. Norberg-King, W. B. Horning II, F. A. Kessler, J. R. Menkedick, T. W. Neiheisel, P. A. Lewis, D. J. Klemm, Q. H. Pickering, E. L. Robinson, J. M. Lazorchak, L. J. Wymer, and R. W. Freyberg. 1989. Short-Term Methods for Estimating the Chronic Toxicity of Effluents and Receiving Waters to Freshwater Organisms, 2nd ed. EPA/600/4-89/001. Cincinnati, Ohio: U.S. Environmental Protection Agency.

Yoccoz, N. G. 1991. Use, overuse, and misuse of significance tests in evolutionary biology and ecology. Bulletin of the Ecological Society of America 72:106–111.

Measures of Environmental Performance and
Ecosystem Condition. 1999. Pp. 217–226.
Washington, DC: National Academy Press.

Measuring Environmental Performance through Comprehensive River Studies

RICHARD STRANG AND LOUIS SAGE

The objective of the Clean Water Act is to restore and maintain the chemical, physical, and biological integrity of the nation's waters. Recently, an intergovernmental task force was established to coordinate and improve the collection and evaluation of monitoring data used in making decisions on water resources. One of the task force's initial activities was to estimate the relative amounts of money spent on water pollution abatement and ambient water-quality monitoring. The task force concluded that for every dollar invested in programs and infrastructure designed to reduce water pollution, less than two-tenths of one cent was spent to monitor the effectiveness of such abatement programs (Intergovernmental Task Force on Monitoring Water Quality, 1992).

Summarizing this state of affairs, the task force stated that "although we have spent more than $500 billion on water pollution abatement since the 1970s, we are currently unable to document adequately the effectiveness of these investments in achieving the objectives of the Clean Water Act and other Federal and State legislation related to water quality" (Intergovernmental Task Force on Monitoring Water Quality, 1992). To continue progress toward achieving the nation's water-quality objectives, more emphasis must be placed on water-quality evaluations and the information such studies can provide about improvements that have been achieved and the issues that remain to be addressed.

APPLICATION OF THE QUALITY MANAGEMENT PROCESS

The quality management process (QMP) is being used effectively today by the manufacturing sector to control and continually improve performance and

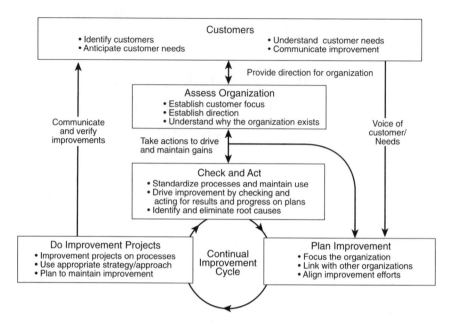

FIGURE 1   The quality management process.

product quality. QMP can serve as a valuable model for addressing water-quality issues. One of the most important features of QMP is the continual improvement cycle, which is intended to accomplish continuous improvements in management systems. The cycle has four phases: planning, doing, checking, and acting. Each phase of this cycle, from designing and implementing projects to checking for improvements, can play an essential role in finding solutions to water-quality issues.

Eastman Chemical, the 1993 Malcolm Baldrige National Quality Award winner, has applied QMP to every aspect of its business, including its water-quality management systems. Figure 1 is a schematic representation of QMP. Eastman organizations that are responsible for water-quality near Eastman facilities have carefully considered the customers for their work. These customers include plant management, the public, employees, the board of directors, stockholders, and state and federal regulatory authorities. Communications between customers and Eastman water-quality organizations are encouraged and help provide the organizations with direction. Conflicting demands often make finding resolutions difficult but do serve to focus attention on important issues.

Based on an understanding of their mission and the demands of their customers, the Eastman water-quality organizations enter the continual improvement cycle and plan improvement projects. These projects range from spill prevention training to waste minimization initiatives and the construction of improved waste treatment facilities. Efforts are made to link with other Eastman organizations and to align with other environmental improvement projects. Projects are then implemented, and customers are informed of the anticipated improvements.

The next step in the cycle is to check for improvements resulting from the water-quality initiatives. For many industries that practice QMP, this check step is accomplished by reviewing readily available measures, such as discharge monitoring reports, that might provide data on improvements in compliance with discharge permits. These measures are also used by Eastman. However, Eastman has gone beyond these measures and now makes periodic river studies an important feature of its water-quality management system.

Eastman has turned to a third party, the Academy of Natural Sciences (ANS) of Philadelphia, to conduct these studies. The academy, founded in 1812, is a world-renowned, nonprofit institution dedicated to environmental research and natural-history education. Over the years, ANS has completed a total of 15 river studies at Eastman's four major plant sites.

These river studies include the collection of chemical, physical, and biological data at locations upstream and downstream of the manufacturing facilities. Special attention is placed on evaluating the resident populations of algae, aquatic plants, noninsect macroinvertebrates, insects, and fish. Through the years, the design of the river assessment has evolved from an emphasis on species richness to a balance between population dynamics and community interactions. These changes are consistent with the early assessments yet allow for more robust program designs that support rigorous statistical analyses.

Results of the river studies have been communicated to Eastman's customers, including the public and regulatory authorities. It is through this communication that the benefits of the studies are realized. Improved understanding of water-quality issues is then incorporated into the next iteration of the continual improvement cycle.

RIVER-STUDY FUNDAMENTALS

ANS river studies are designed to assess the overall health of a river that receives discharges from a facility. Much like a medical checkup, certain indicators are evaluated as the basis for the assessment. The studies focus on components of biological communities that have been shown to be the most sensitive indicators of environmental quality (Schindler, 1987). If abnormalities are detected in one or more of the indicators, additional more-focused studies are recommended. To be properly implemented, the third-party reviews need to be scheduled every 4 to 5 years, depending on the nature of the commercial operation.

The indicators are represented by groups of organisms that reflect different functions in the aquatic community. These groups have varying strengths and weaknesses that, when properly assembled in a program, can be complementary in the overall assessment. This information is supplemented with water-chemistry and physical data collected at the time of the biological collections.

Studies for Eastman Chemical facilities include an assessment of the species composition and relative abundance of algae and diatoms as a measure of the base of the food chain. Insect macroinvertebrates are quantitated to provide information on the biomass of this important fish food resource. The insects are excellent indicators, providing acceptable rigor for statistical testing against a variety of environmental parameters. Noninsect macroinvertebrates, or epibenthic fauna, such as mussels and crawfish support themselves by filtering food particles from the water or by scavenging. Many of these organisms remain in one spot as adults and thus reflect the water-quality of a particular area. The organisms most universally associated with a river, fish, reside at the top of the food chain and thus reflect the health of the community through all the links in that chain. The fish community represents the official report card on the health of the river to the majority of the general public. Because fish are mobile and are attracted to specific habitats, they are less randomly distributed than other aquatic life. Therefore, the presence or absence of a given fish species downstream of a discharge may not be as meaningful as a similar observation related to macroinvertebrates. Fish, however, are an excellent biological group for assessing the effect of water-quality on growth rate. This can be done through inspection of their ear bones, a procedure termed otolith analysis. Fish are also useful for assessing body burdens of river pollutants.

A special challenge in developing a program for a third-party review is to employ state-of-the-science methods while maintaining a database that allows for long-term trend analysis. Such a database allows investigators to examine questions that relate to the rate of change in the composition of the biological community, the age structure of populations, as well as more generic questions relating to local and regional point and nonpoint discharges.

ANS's first work with Eastman was a 1965 study for the Tennessee Eastman Division, which examined a wide variety of habitats to identify as many species as possible in each river reach. Organisms included in this first study were algae, macroinvertebrates, insects, protozoans, and fish. After determining the number of species, the assemblage was sorted according to the pollution tolerance of each group. Based on these groupings, a comparative index was developed from other rivers in the mid-Atlantic region to establish the health of the South Fork Holston River.

During the more recent studies for Tennessee Eastman and Arkansas Eastman, ANS has placed greater emphasis on acquiring quantitative data by determing such things as catch of organisms per unit of effort and the number of specimens per unit of habitat, and through sample replication. These data are then used in statistical tests and to calculate biotic indices. For insect macroinvertebrates, for example, ANS developed a computerized database that includes information on the lowest practicable taxonomic level of each specimen and on functional feeding groups such as predator, shredder, scraper, filterer, and gatherer. With this database, analyses can be conducted to determine taxa richness, abundance, Shannon-Wiener diversity, community evenness, the relative balance of functional feeding groups, and Hilsenhoff's index of pollution tolerance. The latter weighs each taxon's pollution tolerance score by that taxon's proportional abundance in the collection (Hilsenhoff, 1987). Although there has been an effort to extend the database's level of quantification, and the collection effort has been refined and replicated, many elements have remained consistent, allowing for comparison of past and present conditions.

CASE STUDIES

Each of the Eastman facilities faces a different set of water-quality challenges, which are reflected in the corresponding differences in the role of the river studies in the QMP check phase for each facility. These differences are best characterized by comparing the situations at the Eastman divisions in Kingsport, Tenn., and Batesville, Ark.

Tennessee Eastman Division

Tennessee Eastman Division began operations in 1920 and has grown to become one of the largest manufacturing facilities in the United States. The plant occupies over 1,000 acres, has over 400 manufacturing buildings, and employs some 12,000 people.

Water-quality issues in the vicinity of the plant are extremely complex. The watershed for the South Fork Holston River, which flows past the facility, is regulated by a series of five dams. The nearest of these, Fort Patrick Henry Dam, is less than 3 miles upstream of the facility. At one time, there were 42 point-source discharges along the 5-mile reach of river that flows through the Kingsport community (Tennessee Department of Public Health, 1977). Today, the major discharges include cooling water and treated process waste water from Tennessee Eastman, as well as other discharges from a munitions manufacturer, a paper manufacturing plant, and a domestic waste-water treatment facility. All of these discharges occur along a 2-mile reach of the river.

In 1970, the South Fork Holston River was one of the four most polluted major Tennessee rivers (Thackston et al., 1990). According to Thackston et al., at this time, biochemical oxygen demand (BOD) loadings from all point sources in Kingsport were as high as 137,000 lb/day.

During the first ANS study for Tennessee Eastman Division in 1965, it became apparent that specific determinations of the causes of observed water-quality problems were impossible because of the proximity of the numerous point-source discharges and the added complication of river flow regulation immediately upstream of Kingsport. Because of these physical constraints, ANS studies have focused on documenting changes in water-quality brought about by investments in water protection by area facilities responsible for discharges into the river.

Thackston et al. 1990 assert that improvements in water-quality downstream of Kingsport are a success story for Tennessee. ANS studies conducted in 1965, 1974, 1977, 1980, and 1990 confirm this claim, with comprehensive data on the increases in species diversity from the improved water-quality conditions. By 1990, BOD was reduced to less than 6,000 lb/day.

Arkansas Eastman Division

Construction of Arkansas Eastman began in 1975, and the facility was in full operation in early 1977. Today, the manufacturing operation occupies 40 acres and employs approximately 700 people. The plant's design was conceived at a time when attitudes toward the environment were heavily influenced by events such as Earth Day, the creation of the U.S. Environmental Protection Agency, and the passage of the Clean Water Act. Lessons learned at Eastman's facility in Tennessee on the protection of water-quality were incorporated into the design of the Arkansas plant.

Arkansas Eastman is located in a rural area; the nearest point source on the White River is over 9 miles upstream. The facility is equipped with an activated-sludge waste-water treatment plant and an incinerator for the combustion of concentrated wastes. All runoff from manufacturing areas is collected and routed through a large holding basin. If an accident were to happen, chemical spills or deluge water could be captured and prevented from reaching the river.

ANS conducted the first of its river studies for Arkansas Eastman in 1974 and 1976, before the facility began to operate. These studies were designed to provide baseline data. Later studies, conducted in 1980 and 1991, involved data collection upstream and downstream of the Arkansas Eastman discharge. Unlike the work in Tennessee, which sought to document improvements in water-quality, the studies in Arkansas were intended to evaluate critically any changes in water-quality that could be attributed to Arkansas Eastman's presence. To date, the studies reflect positively on the water-quality management system at Arkansas Eastman. The data indicate no adverse impacts due to plant operations.

RESULTS AND DISCUSSION

The ANS studies provide both qualitative and quantitative data on the health of aquatic communities near Eastman facilities. Information on species diversity has been collected throughout the 25-year history of the program. This information is particularly useful for comparing trends in water-quality with changes in management systems.

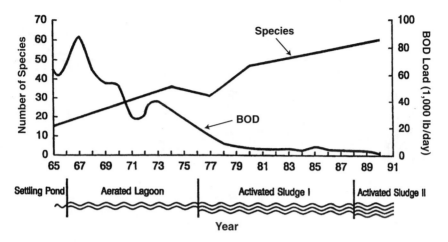

FIGURE 2   Tennessee Eastman Division waste-water treatment capability and corresponding Eastman BOD load to the South Fork Holston River compared with the numbers
of species of invertebrates and fish at a location downstream of Kingsport, Tennessee,
1965–1990.

Tennessee Eastman Division

Historical information on Tennessee Eastman Division's waste-water treatment capability and BOD loading to the South Fork Holston River is provided in
Figure 2.   In 1966, Eastman converted from the use of simple settling ponds to
biological waste-water treatment with aerated lagoons.   From the late 1960s to
the early 1970s, company efforts focused on minimizing waste streams and eliminating waterborne discharges.   Combustion units with energy-recovery boilers
were constructed to incinerate wastes that were isolated from manufacturing processes.   By the early 1970s, these activities resulted in a 65 percent reduction in
the Tennessee Eastman BOD loading to the South Fork Holston River.

Data on aquatic invertebrates and fish collected at a location downstream of
all Kingsport area point-source discharges, before and after this BOD load reduction, showed a 140 percent increase in the total number of species of invertebrates
and fish, from 15 species in 1965 to 36 species in 1974.

In 1976, Tennessee Eastman Division again improved its waste-water treatment capability by installing an activated-sludge treatment system.   This change,
combined with continued emphasis on waste minimization, resulted in additional
improvements that amounted to a 96 percent reduction in Tennessee Eastman's
BOD loading between 1967 and the early 1980s.   Other point sources in the
Kingsport area also achieved load reductions during this period.   As a result,
there was a 210 percent increase in the total number of species of invertebrates
and fish, from 15 species in 1965 to 47 species in 1980.

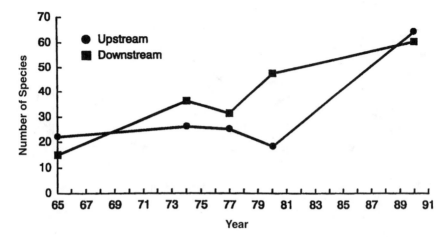

FIGURE 3   Comparison of the total number of fish and invertebrate species at locations upstream and downstream of the Kingsport area point-source discharges, 1965–1990.

In 1988, Tennessee Eastman Division began operating a new $90 million advanced activated-sludge waste-water treatment plant. Subsequently, compared with 1967 levels, the Tennessee Eastman BOD load to the river was reduced by 99 percent. This new technology produced only a 3 percent change in load reduction compared with the activated-sludge system used throughout the 1980s. ANS scientists were surprised to find dramatic increases in the number of insect and fish species, from 47 species in 1980 to 60 species in 1990, which could not be accounted for by the relatively small reduction in BOD. Altogether, there was a 300 percent increase in the number of fish and invertebrate species between 1965 and 1990.

After verifying that collection techniques and river conditions during the studies were similar enough to make the data comparable, the researchers focused on possible changes upstream of the point-source discharges to explain the observed changes. Figure 3 compares data on total fish and invertebrate species for locations upstream and downstream of Kingsport during the 1965–1990 study period. Increases in species diversity downstream of Kingsport from 1965 to 1980 can be attributed to BOD load reductions resulting from cleaner point-source discharges. Little change took place during this period at the upstream station, which is located between Tennessee Eastman Division property and Fort Patrick Henry Dam. However, increased species diversity observed at this location in the 1990 study indicates improved conditions that are not associated with the Kingsport area point-source discharges. The change in conditions at the upstream station apparently also resulted in improvements at locations farther downstream.

Arkansas Eastman Division

ANS conducted baseline studies of the White River in 1974 and 1976 before manufacturing operations started at Arkansas Eastman Division. These studies were aimed at generating data for use in comparison with future studies and to assess the natural variability in resident populations of aquatic life.

The 1980 and 1991 ANS studies used three types of analysis to assess the potential effects of Eastman operations on the White River: comparisons of the 1980 and 1991 data with baseline data from 1974 and 1976; comparisons of upstream and downstream data; and comparisons of left-bank and right-bank data. These analyses were designed to show evidence of impact due to discharges by Arkansas Eastman, for example if the biological parameters at sampling locations or over time changed in a pattern consistent with exposure to point-source pollution.

Table 1 shows the number of noninsect macroinvertebrate, insect, and fish species collected at stations immediately upstream and downstream of the Arkansas Eastman discharges to the White River. These and other data collected during the studies on species richness and abundance show differences among the sampling locations, but the observed patterns are not consistent with a negative effect from the Eastman facility. All indications are that the water-quality management systems in place at Arkansas Eastman are successfully protecting the White River.

SUMMARY

ANS river studies have made significant contributions to Eastman's "plan, do, check, and act" approach to improving and maintaining water-quality management systems. The case studies for Tennessee Eastman and Arkansas Eastman illustrate the importance of developing and understanding the effects that water-quality initiatives have on the resources they are meant to protect. For Arkansas Eastman management, the studies provide reassurance that the systems in place to protect the White River are functioning as planned. At Tennessee Eastman, the

TABLE 1   Comparison of Numbers of Aquatic Species Upstream (up) and Downstream (down) of Arkansas Eastman

Taxon	1974		1976		1980		1991	
	Up	Down	Up	Down	Up	Down	Up	Down
Macroinvertebrates	4	3	14	9	8	8	16	17
Insects	43	45	30	24	37	30	47	58
Fish	20	16	—	—	27	34	42	48

studies document the improvements in water-quality brought about by years of investment in point-source controls. However, the Tennessee studies also indicate that investment in such controls and the corresponding improvement in water-quality can reach a point of diminishing returns. Further improvements in water-quality, in these situations, can only be achieved when the remaining issues, such as nonpoint-source pollution, are understood and addressed.

REFERENCES

Hilsenhoff, W. L. 1987. An improved biotic index of organic stream pollution. Great Lakes Entomology 20:31–39.

Intergovernmental Task Force on Monitoring Water Quality. 1992. Ambient Water Quality Monitoring in the United States: First Year Review, Evaluation and Recommendations. Report to the Office of Management and Budget. Washington, D.C.: U.S. Government Printing Office.

Schindler, D. W. 1987. Detecting ecosystem responses to anthropogenic stress. Canadian Journal of Fisheries and Aquatic Sciences 44 (Suppl. 1):6–25.

Tennessee Department of Public Health. 1977. Water Quality Management Plan for the Holston River Basin. Nashville, Tenn.: Division of Water Quality Control.

Thackston, E. L., W. R. Miller, D. Durham, and L. R. Richardson. 1990. Tennessee Environmental Quality Index 1970–1990. Nashville, Tenn.: Tennessee Conservation League Report.

Measures of Environmental Performance and
Ecosystem Condition. 1999. Pp. 227–259.
Washington, DC: National Academy Press.

Biological Criteria for
Water Resource Management

CHRIS O. YODER AND EDWARD T. RANKIN

This paper has two goals: to describe a framework for developing biological criteria for water-quality assessment and to suggest what roles biological indicators should have in water resource management and policy. A principal objective of the Clean Water Act is to restore and maintain the physical, chemical, and biological integrity of the nation's surface waters (Clean Water Act Section 101[a][2]). Although this goal is fundamentally ecological, regulatory agencies have attempted to reach it by measuring chemical and physical, but not biological, variables (Karr et al., 1986). The rationale for this choice is well known; the chemical water-quality criteria developed through laboratory toxicity tests on selected aquatic organisms serve as surrogates for the ecologically based goals of the Clean Water Act.

The presumption that improvements in chemical water quality will restore biological integrity has come into question during the past 20 years. The traditional chemical water-quality approach may give an impression of empirical validity and legal defensibility, but it does not directly measure the ecological health and well-being of surface water resources. Nor does it comprehensively address the definition of pollution in Section 502 of the Clean Water Act: ". . . man-made or man-induced alteration of the chemical, physical, biological or radiological integrity of water . . ." (Karr, 1991). State and federal programs have become focused on the objective of controlling point-source discharges of chemicals. However, a growing number of professionals and various stakeholders have become convinced that an attack on point sources of toxins alone will not fully achieve the Clean Water Act's goal related to surface waters. In addition to an overemphasis on point sources and toxins, many state and federal programs also suffer from

227

- an overreliance on prescriptive approaches to management and regulation;
- a tendency to rely on anecdotal information; and
- an emphasis on administrative activities that frequently results in a skewed allocation of resources among different programs.

Finally, attainment of national goals is also hindered by a lack of consistency in the environmental statistics reported by different states. One result is that in some cases, well-intentioned but basically flawed management strategies have increased environmental degradation as shown, for example, by the Ohio Environmental Protection Agency (1992). Fortunately, the U.S. Environmental Protection Agency (EPA) and others are becoming increasingly aware of these shortcomings and have initiated efforts such as the environmental indicators initiative and the Intergovernmental Task Force on Monitoring Water Quality (1992, 1993, 1995).

Major Factors That Determine Water Resource Integrity

Beyond chemical contaminants, multiple factors are responsible for the continuing decline of surface water resources in Ohio (Ohio Environmental Protection Agency, 1995) and the United States (Benke, 1990; Judy et al., 1984). These include the modification and destruction of riparian habitat, sedimentation of bottom substrates, and alteration of natural flow regimes. Because biological integrity is affected by many factors, controlling chemicals alone does not assure its protection or restoration (Figure 1). We need a broader focus on the entire water resource if we are to successfully reverse the decline in the overall quality of the nation's waters. Therefore, ecological concepts and biological criteria must be further incorporated into the management of surface water resources.

Disparities in the Use of Indicators

Although a growing number of states and organizations rely primarily on biological indicators to assess the condition of their water resources, others choose to emphasize chemical and physical indicators. The following examples demonstrate the inherent risks of relying solely on these indicators.

Out of 645 stream and river segments analyzed in Ohio, biological indicators revealed impairment in 49.8 percent of the segments where chemical indicators detected none (Ohio Environmental Protection Agency, 1990; Rankin and Yoder, 1990a). The converse was true for only 2.8 percent of stream segments. The remarkable discrepancy between biological and chemical assessments is due to fundamental differences in what they measure. Biological communities respond to a wide variety of chemical, physical, and biological factors. Thus, biological indicators are able to detect a wider range of environmental disturbances than can measures of chemical water-quality alone.

Another example is the proportion of waters that various states reported were

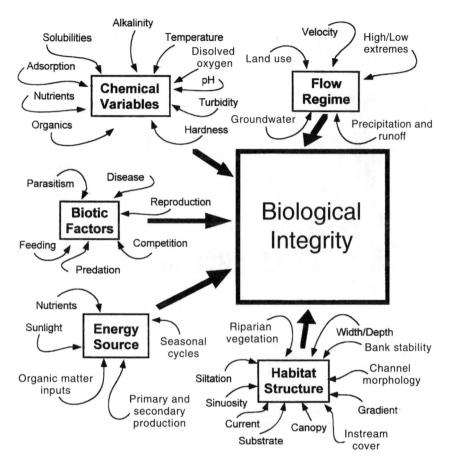

FIGURE 1   Five principal factors that influence and determine the integrity of surface water resources. SOURCE: Modified from Karr et al. (1986).

impaired due to habitat degradation (Figure 2). Twenty-five of 58 states and territories that report such statistics claimed that *zero* miles of rivers and streams had been negatively affected by the modification of habitat. Of the states that did report such impairment, only 15 reported an effect on more than 100 miles. These statistics are difficult to believe given the pervasive nature of well-documented practices that modify habitat, such as flood control, impoundments, agriculture and forestry, resource extraction, and urban development (Benke, 1990; Judy et al., 1984). The wide variation in state statistics is probably due to the use of different indicators and programmatic biases toward the control of toxic chemicals and point-source discharges (Ohio Environmental Protection Agency, 1990; Rankin and Yoder, 1990a).

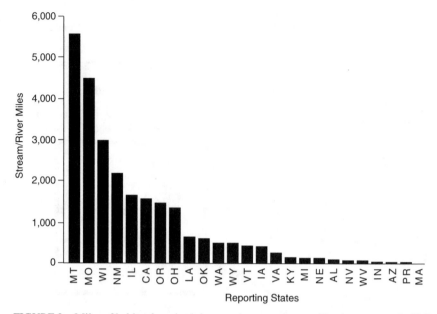

FIGURE 2   Miles of habitat-impaired rivers and streams reported by the states to the U.S. Environmental Protection Agency. SOURCE: United States Environmental Protection Agency (1994). NOTE: Twenty-five states reported no miles of aquatic-life use impairment associated with habitat degradation.

Ohio's use of quantitative biological criteria had some additional ramifications that affect the statistics related to the Clean Water Act 305[b]. For example, the proportion of stream miles that failed to attain standards increased from 9 percent in 1986 (based on a mix of chemical water-quality and qualitative biological assessments) to 44 percent in 1988, primarily because quantitative biological criteria were included in the assessment process beginning in 1988 (Ohio Environmental Protection Agency, 1988). The nearly fivefold increase in nonattainment illustrates the significant differences that can exist between states that use different assessment methods, especially whether or not biological assessments are included.

These examples demonstrate that relying on chemical water-quality information alone is apt to result in underestimates of environmental degradation. Underestimates are especially likely when assessing watershed-level effects. This is because the interaction of aquatic and riparian habitats, land use, and nutrient dynamics is particularly difficult to measure and characterize without using robust biological assessment tools and indicators. Ironically, much of the concern expressed about using biological criteria has been over the risk of failing to detect water-quality impairment. This concern seems misplaced in light of the preceding examples.

The Complementary Roles of Different Indicators

The EPA Environmental Monitoring and Assessment Program (United States Environmental Protection Agency, 1991) distinguishes environmental indicators on the basis of whether they best measure stresses, exposures, or responses. Stressor indicators identify activities that have an impact on the environment. These include land-use changes and discharges of pollutants. Exposure indicators identify components of the environment that have been subjected to a substance or activity that could potentially change the environment directly or indirectly. Response indicators detect the status of a particular resource component, usually biological, in relation to external stresses and exposure to those stresses. Particular indicators function most appropriately in one of the three indicator categories, although they may double as a secondary evaluator of another indicator class. For example, chemical measures generally function best as exposure indicators but may indirectly provide insights to response. Biological measures are inherently response oriented and may or may not provide more than qualitative insights to exposure and stressors.

These comparisons of chemical and biological indicators illustrate a national problem: the inappropriate use of stressor and exposure indicators as substitutes for response indicators. States that do not have well-developed bioassessment programs still must report on the status of their waters to EPA on a biennial basis. Thus, they are forced to use whatever information is available. Usually, the readily available information is in the form of stressor or exposure indicators. In attempting to resolve the obvious inconsistencies in measuring the condition of aquatic resources, a fundamental step is to recognize and establish appropriate roles for the different chemical, physical, and biological indicators. An accurate portrayal of the condition of the nation's surface waters depends on the use of suites of these indicators, each in their appropriate role as stressor, exposure, or response indicators.

DEVELOPMENT OF BIOLOGICAL CRITERIA

Underlying Theory and Concepts

Without a sound theoretical basis, it would be difficult if not impossible to develop biological criteria and meaningful measures of ecological condition. Obvious ecological degradation such as fish kills stimulated the landmark environmental legislation of the past 2 decades, but that biological focus was lost in the quest for easily measurable water-quality indicators (Karr, 1991). The biological integrity provision of the Clean Water Act, which was initially difficult to specify in practice (Ballentine and Guarria, 1975), was eventually defined by Karr and Dudley (1981) as ". . . the ability of an aquatic ecosystem to support and maintain a balanced, integrated, adaptive community of organisms having a species composition, diversity, and functional organization comparable to that of the natural habitats of a region." It was this definition that provided the theoretical underpinnings for developing a framework within which quantitative biological criteria could be derived. The essential concepts of how to measure and define biological performance, natural habitats, and regional variability were each dealt with through a number of key research projects in the early 1980s and together provided the framework and tools needed to derive biological criteria.

Given the above definition, the eventual attainment of the Clean Water Act goal of biological integrity requires much more than merely achieving a high level of species diversity, numbers, and/or biomass. In fact, in some situations, increases in any one of these parameters can be a sign of degradation. Managers must also strive for more than the health and well-being of certain target species. The conservation of individual species, although necessary, is not sufficient for ensuring biological integrity. Conservation policy should promote management practices that maintain integrity, prevent endangerment, and enhance the recovery of species and ecosystems (Angermier and Williams, 1993). Water-resource management must include in its goal the objective of achieving self-sustaining, functionally healthy aquatic ecosystems. Achieving this state will foster other ecological goals as well because functionally healthy communities include the elements of biodiversity and rare species inherent to the more narrowly focused management efforts. We believe that biocriteria can play an important role in meeting these challenges.

Understanding Biological Integrity: A Prerequisite to Biological Criteria

The term biological integrity originated in Section 101[a][2] of the Federal Water Pollution Control Act amendments of 1972 and has remained a part of the subsequent reauthorizations. Early attempts to define biological integrity in ways that could be used to measure attainment of the legislative goals were inconclusive. One of the better known of these efforts failed to produce a consensus definition or framework for determining biological integrity (Ballentine and Guarria, 1975). Biological integrity was considered relative to conditions that existed prior to European settlement; the protection and propagation of balanced, indigenous populations; and ecosystems that are unperturbed by human activities. These criteria (especially the first and last) could be construed as referring to a pristine condition that exists in only a few, if any, ecosystems in the United States. Subsequently, an EPA-sponsored work group concluded that it is difficult and perhaps impractical to precisely define and assess biological integrity, when it is viewed as a pristine condition (Gakstatter et al., 1981). Rather, the pristine vision of biological integrity became a conceptual goal of pollution abatement efforts, although it may never be fully realized in many parts of the United States given current, past, and future uses of surface waters.

More recently, efforts to construct a workable, practical definition of biological integrity have provided the supporting theory needed to develop standardized measurement techniques and criteria to determine whether efforts are complying with that goal. Biological integrity is now defined as ". . . the ability of an aquatic ecosystem to support and maintain a balanced, integrated, adaptive community of organisms having a species composition, diversity, and functional organization comparable to that of the natural habitats of a region" (Karr and Dudley, 1981). This is a workable definition that directly alludes to the measurable characteristics of biological community structure and function found in the least-impacted habitats of a region. This definition and its underlying ecological theory provide the basis for developing quantitative biological criteria based on conditions at regional reference sites. The EPA adopted a facsimile of this definition in their biological criteria national program guidance (United States Environmental Protection Agency, 1990).

The emerging issue of biodiversity should not be equated with biological integrity, even though the two concepts share many attributes (Karr, 1991). They differ in that biodiversity is primarily focused on ecosystem elements (i.e., genetic diversity, populations, bioreserves, etc.), whereas biological integrity includes these elements but also encompasses ecosystem processes (i.e., nutrient cycles, trophic interactions, speciation, etc.). The often-cited ecosystem approach to environmental management (e.g., Great Lakes Water Quality Initiative) can be even more restricted to dealing with elements that are not direct ecological parameters (i.e., chemical water-quality surrogates). Both the biodiversity and the ecosystem approaches would benefit by including the concept of biological integrity to improve the chances that each effort would succeed and assure that environmental problems are addressed from an ecological perspective.

New Multimetric Biological-Community Evaluation Mechanisms

A variety of quantitative indices for assessing biological data have been developed in the past 20 years. These indices represent significant advances because they use biological information for resource characterizations and for determining the attainment of environmental goals. Examples include the Index of Biotic Integrity, as originally developed by Karr (1981) and modified by many others (Leonard and Orth, 1986; Miller et al., 1988; Ohio Environmental Protection Agency, 1987b; Steedman, 1988); the Index of Well-Being (Gammon, 1976; Gammon et al., 1981); the Invertebrate Community Index (DeShon, 1995; Ohio Environmental Protection Agency, 1987b); the EPA Rapid Bioassessment Protocols for macroinvertebrate assemblages (Plafkin et al., 1989); and the Benthic Index of Biotic Integrity (Kerans and Karr, 1992).

Although quantitative biological indices have been criticized for potentially oversimplifying complex ecological processes (Suter, 1993), raw data must be distilled to be interpretable. Multimetric evaluation mechanisms extract ecologically relevant information from complex biological data while preserving the opportunity to analyze such data on a multivariate basis. Several features of these multimetric indices minimize the problem of data variability. Variability is first controlled by specifying standardized methods and procedures (e.g., Ohio Environmental Protection Agency, 1989b) and providing iterative training exercises and supervision to implement them. Second, variability is in effect "compressed" through the application of multimetric evaluation mechanisms (e.g., Index of Biotic Integrity, Invertebrate Community Index), which reduce raw measurements into discrete, calibrated scoring categories. Last, variability is partitioned according to background factors that determine ecological potential (e.g., ecoregions), a process that results in a graduated set of criteria based on regional potential. The results are evaluation mechanisms, such as the Index of Biotic Integrity and the Invertebrate Community Index, that have acceptably low replicate variability (Davis and Lubin, 1989; Fore et al., 1993; Rankin and Yoder, 1990b; Stevens and Szczytko, 1990).

Multimetric indices have been criticized as representing a loss of rich information because the data are reduced to a single index value. However, this presumes that the supporting data are never viewed or examined beyond the calculation of the index itself. These criticisms are without foundation: The need to examine subcomponents of the indices and even the raw data is implicit throughout the biocriteria process. Theoretically sound quantitative measures, as opposed to raw data, are clearly necessary throughout the process of environmental management. Although interpretation of raw data by qualified biologists will always be necessary, it is not realistic to expect that their qualitative judgment alone will be an acceptable substitute for a more empirical process. Fortunately, biological judgment can be incorporated into structured frameworks. These include the frameworks developed by the state of Maine using multivariate techniques (Davies et al., 1993), the state of Florida using a multimetric approach (Barbour et al., 1996), the EPA Rapid Bioassessment Protocols (Plafkin et al., 1989), and the regional reference-site approach (Hughes et al., 1986; Ohio Environmental Protection Agency, 1987b, 1989a; Yoder and Rankin, 1995a). Simply stated, multimetric indices can satisfy the demand for a straightforward numerical evaluation that expresses a relative value of aquatic community health and well-being and allows program managers (who are frequently nonscientists) to "visualize" relative levels of biological integrity. These measures also provide a means to establish quantitative biological criteria.

Karr's Index of Biotic Integrity Modified by the Ohio Environmental Protection Agency

The Index of Biotic Integrity originally proposed by Karr (1981) and later refined by Karr et al. (1986) and others incorporates 12 metrics (Table 1). These fall within four broad groupings: species richness and composition, trophic composition, environmental tolerance, and fish abundance and condition. Although no single metric consistently functions across an entire environmental gradient

TABLE 1   Index of Biotic Integrity Metrics Used by the Ohio EPA to Evaluate Headwater Sites, Wading Sites, and Boat Sites

Metric	Headwater Sites[a,b]	Wading Sites[b]	Boat Sites[c]
1. Number of native species[d]	X	X	X
2. Number of darter species		X	
Number of darter and sculpin species	X		
Percent round-bodied suckers[e]			X
3. Number of sunfish species[f]		X	X
Number of headwater species[g]	X		
4. Number of sucker species		X	X
Number of minnow species	X		
5. Number of intolerant species		X	X
Number of sensitive species[h]	X		
6. Percent green sunfish			
Percent tolerant species	X	X	X
7. Percent omnivores	X	X	X
8. Percent insectivorous cyprinids			
Percent insectivores	X	X	X
9. Percent top carnivores		X	X
Percent pioneering species [i]	X		
10. Number of individuals			
Number of individuals (less tolerants)[j]	X	X	X
11. Percent hybrids			
Percent simple lithophils		X	X
Number of simple lithophils	X		
12. Percent diseased individuals			
Percent DELT anomalies[k]	X	X	X

[a]Sites with drainage areas <20 sq. mi.

[b]Sampled with wading electrofishing methods.

[c]Sampled with boat electrofishing methods.

[d]Excludes all exotic and introduced species.

[e]Includes all species of the genera *Moxostoma, Hypentelium, Minytrema,* and *Ericymba,* and excludes *Catostomus commersoni.*

[f]Includes only *Lepomis* species.

[g]Species designated as permanent residents of headwaters streams.

[h]Includes species designated as intolerant and moderately intolerant (Ohio Environmental Protection Agency, 1987b).

[i]Species designated as frequent and predominant inhabitants of temporal habitats in headwater streams.

[j]Excludes all species designated as tolerant, hybrids, and non-native species.

[k]Includes only animals with deformities (D), eroded (E) fins or barbels, lesions (L), or tumors (T).

NOTE: This table lists the original Index of Biotic Integrity metrics of Karr (1981). In cases where the Ohio EPA uses modifications of the original metric, the modifications appear below the original metrics.

and for all types of impacts, their aggregation in the Index of Biotic Integrity provides sufficient overlap and redundancy to yield a consistent and sensitive measure of biological integrity (Angermier and Karr, 1986). The index is a quantitative, ordinal, if not linear, measure that responds in an intuitively correct manner to known environmental gradients (Steedman, 1988). When incorporated with mapping, monitoring, and modeling information, the Index of Biotic Integrity is valuable for determining management and restoration requirements for warm-water streams (Bennet et al., 1993; Steedman, 1988). As an aggregation of community information, the Index of Biotic Integrity and its facsimiles provide a way to organize complex data and reduce it to a form that permits interpretation and comparisons with communities whose condition is known. Although the Index of Biotic Integrity incorporates elements of professional judgment, it also provides the basis for quantitative criteria for determining what constitutes exceptional, good, fair, poor, and very poor conditions.

The process of tailoring the Index of Biotic Integrity to regional conditions represents an important example of the use of biological judgment in biological criteria (Miller et al., 1988). Streams and rivers occur in many sizes throughout Ohio. They contain different fish assemblages and must be sampled by different methods. Thus, the Ohio EPA needed to modify Karr's original index to apply it to these different stream sizes and adjust it to account for biases induced by different sampling gear. The three different modified Indices of Biotic Integrity that were developed for Ohio rivers and streams (Table 1; Ohio Environmental Protection Agency, 1987b) are: 1) a headwaters index for application to headwater streams (locations with a drainage area <20 mi^2); 2) a wading-site index for application to stream locations with watersheds >20 mi^2 sampled with wading methods; and 3) a boat-site index for locations sampled from boats. All modifications follow the guidance on metric modification provided by Karr et al. (1986).

Although the Index of Biotic Integrity has worked well in Ohio and in many parts of the United States (Miller et al., 1988), Canada (Steedman, 1988), and Europe (Oberdorff and Hughes, 1992), problems have been encountered in semi-arid western U.S. drainages (Bramblett and Fausch, 1991) and cold-water streams in northern states (Lyons, 1992). In both cases, the characteristics of the fauna differ from the presumptions made in the original index. Lyons (1992) was initially confounded by the degradation of cold-water streams in Wisconsin because it resulted in an increased diversity of fish species, a change that is counted as an improvement in water quality by the original Index of Biotic Integrity. He overcame these problems and constructed an index tailored to cold-water streams (Lyons and Wang, 1996). Bramblett and Fausch (1991) encountered difficulties due to a lack of pollution-intolerant species in the harsh and highly variable hydrological conditions of the Arkansas River basin in Colorado. These examples emphasize the need to consider the inherent characteristics of the regional fauna when developing multimetric biological assessment tools.

Because surface water resources naturally vary across the nation, nationally uniform biological criteria are neither feasible nor desirable. However, it is desirable to have a national framework based on the concepts of the regional reference-site approach (Hughes et al., 1986), which will promote national consistency between state bioassessments. The key is to describe the framework within a common national goal, such as the maintenance and restoration of biological integrity.

Initial Considerations

A number of fundamental decisions need to be made prior to adopting a set of biological monitoring methods. This is a critical juncture in the process because bad decisions will reduce the effectiveness of the efforts well into the future. These vital initial choices include which sampling methods to use, when to sample, which organisms to monitor, which parameters to measure, and which level of taxonomic precision to use. If any axiom applies, it is: When in doubt, take more measurements than seem necessary at the time because uncollected information cannot be retrieved later. Parameters that require little or no extra effort to acquire should be included until the evidence shows they are unnecessary. One example in Ohio is external anomalies on fish. We decided to record this information even though it was not apparent how it would be useful. This measure has proved to be one of our most valuable assessment tools. For macroinvertebrates, the decision to identify midges to the genus and species level was also fortuitous given the value of this group in diagnosing impairments. Of course, samples could always be archived for later reinspection, but the logistical burdens that this entails are undesirable.

Another important consideration is to assure that qualified and regionally experienced staff do the fieldwork. In ecological assessment, like many other professions, the most skilled and experienced individuals are sought to direct, manage, and supervise. However, biological field assessment requires an equivalent level of expertise in the field because many of the critical pieces of information are recorded and, to a degree, interpreted in the field. There is simply no substitute for the intangibles gained by direct observations in the field. This is not a job to be left to technicians. The staff who perform the fieldwork should also plan that work, process the data, interpret the results, and write reports. Such staff, particularly the more experienced individuals, also contribute to policy development.

Water-Quality Standards: Designated Uses and Criteria

The Ohio water-quality standards (Ohio Administrative Code 3745-1) consist of designated habitat classifications. Chemical, physical, and biological criteria are specific to each classification and are designed to be consistent with the goals specified by each classification. Protection and restoration requirements are a function of habitat classification. The Ohio water-quality standards describe five different habitat classifications. The general intent of each with respect to biological criteria are

- Warm-water habitat—This designation defines the most commonly occurring warm-water assemblages of aquatic organisms in Ohio rivers and streams and represents the principal *restoration* target for the majority of water resource management efforts in Ohio. Biological criteria for warm-water habitat are tailored to the five different ecoregions within Ohio.
- Exceptional warm-water habitat—This designation is reserved for waters that support "unusual and exceptional" assemblages of aquatic organisms with exceptionally high species richness. Pollution-intolerant species are commonly present, as are significant populations of rare, threatened, or endangered species. This designation represents a *protection* goal for Ohio's best water resources. Biological criteria for exceptional warm-water habitat are the same for each ecoregion.
- Cold-water habitat—This designation is used for sites that support assemblages of cold-water organisms and sites that are stocked with salmonids for put-and-take fishery purposes. No specific biological criteria have been developed for this habitat classification. The warm-water habitat biocriteria are used to evaluate cold-water habitat sites.
- Modified warm-water habitat—This designation applies to streams and rivers that have been subjected to extensive, persistent, and essentially permanent hydromodifications that are permitted by state or federal law. The warm-water habitat biocriteria are not attainable at these sites. Species present are generally tolerant of low dissolved oxygen, silt, nutrient enrichment, and poor quality habitat. The biological criteria for modified warm-water habitat were derived from a particular set of habitat-modified reference sites and are tailored to the five ecoregions and three major modification types: channelization, run-of-river impoundments, and sites with extensive sedimentation due to nonacidic mine drainage.
- Limited resource water—This designation applies to small streams (usually <3-mi^2 drainage area) and other water courses that have been altered to the extent that they can support no functional assemblage of aquatic life. Such waterways generally include small streams in extensively urbanized areas, those that lie in watersheds with extensive drainage modifications, those that completely lack water on a recurring annual basis (i.e., true ephemeral streams), and other irretrievably altered waterways. No formal biological criteria have been established for limited resource water because no organized community is to be expected.

Chemical, physical, and/or biological criteria are assigned to each habitat classification according to the broad ecological goals defined by each. This system constitutes a "tiered" approach in that different levels of protection are designated for each habitat classification. The hierarchy is especially apparent for parameters such as dissolved oxygen, ammonia-nitrogen, and biological criteria. For other common parameters such as heavy metals, the technology to develop an equally graduated set of criteria has been lacking; thus, the same water-quality criteria may apply to two or three different habitat classifications. The concepts inherent to the system of habitat classifications also reflect the necessity of reconciling ideals, such as the restoration of biological integrity everywhere, with the lasting effects of two centuries of intensive human use of land and water resources.

Framework for Deriving Quantitative Biological Criteria: Ohio Example

The Ohio EPA adopted quantitative biological criteria in the Ohio water-quality standards regulations in 1990, following a 7-year development process based on a 10-year database. These criteria are based on measurable characteristics of fish and macroinvertebrate communities such as species richness, key taxonomic groupings, functional guilds, environmental tolerances, and organism condition (Ohio Environmental Protection Agency, 1987a,b, 1989a,b; Yoder, 1989; Yoder and Rankin, 1995a).

The resulting quantitative biological criteria represent the degree of biological integrity that can reasonably be expected given present background conditions. Although the process does not purport to attempt to define "pristine," pre-Columbian conditions, the design framework includes a provision to change the biocriteria in response to any future improvements in conditions at reference sites. If, however, general ecological decline continues, today's reference conditions will provide an important link from the past to the future. Regardless, this process does not lock in a particular baseline reference condition. Instead, it establishes a realistic range of desired states of ecological health and well-being against which we can evaluate contemporary environmental management and restoration efforts.

Quantitative biological criteria for Ohio's rivers and streams were derived using data from more than 350 reference sites that typify the "least impacted" condition within each Ohio ecoregion (Ohio Environmental Protection Agency, 1987b, 1989a; Yoder and Rankin, 1995a). This information was then used within the existing framework of habitat classifications in the Ohio water-quality standards to establish regional performance expectations for biological communities.

The framework included the following major steps:

- selection of indicator organism groups;
- establishment of standardized field sampling, laboratory, and analytical methods;
- selection and sampling of least-impacted reference sites;
- calibration of each metric in accordance with the methods described by Karr et al. (1986);

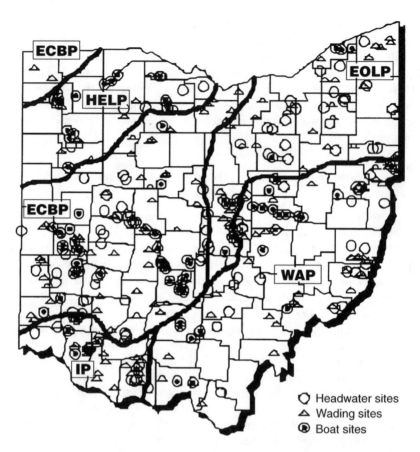

FIGURE 3A   Map showing regional reference sites for calibration of the Index of Biotic
Integrity in Ohio. All of the locations indicated on the map are reference sites. Heavy
black lines show boundaries between the five ecoregions: Huron/Erie Lake Plain (HELP),
Interior Plateau (IP), Eastern-Ontario Lake Plain (EOLP), Western Allegheny Plateau
(WAP), and Eastern Corn Belt Plains (ECBP).

- selection of numeric biocriteria based on the attributes specified by the
various habitat classifications;
- resampling of reference sites (10 percent of sites sampled each year); and
- a planned periodic (i.e., once per decade) review of the calibration and deri-
vation process and, if necessary, adjustment of the multimetric indices, numeric
biological criteria, or both if justified by resampling results from reference sites.

The following example describes the calibration of the Index of Biotic Integrity modification for wading sites. Regional reference sites were first selected
and sampled (Figure 3a) according to procedures outlined by the Ohio Environ-

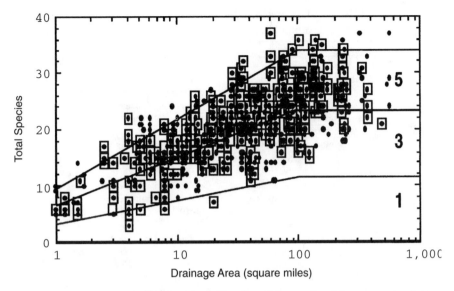

FIGURE 3B   An example showing the calibration of the species richness metric of the Index of Biotic Integrity for wading and headwater site types. The solid lines divide the figure into regions with Index of Biotic Integrity species richness scores of 1, 3, and 5.

mental Protection Agency (1987b, 1989b). The reference biological database was then used to calibrate the Index of Biotic Integrity metrics (Figure 3b). Three different modifications of the Index of Biotic Integrity were then constructed, one each for headwaters, wading (Table 2), and boat sites. The reference-site results were then used to establish the quantitative biological criteria. Notched box-and-whisker plots portray the results for each biological index by ecoregion (Figure 4). These plots convey sample size, medians, ranges with outliers, and 25th and 75th percentiles. Unlike means and standard errors, box plots do not assume a particular distribution of the data. Furthermore, outliers do not exert an undue influence, as they can on means and standard errors. In establishing biological criteria for a particular area or ecoregion, we attempt to represent the typical biological community, not the outliers. Outliers can be dealt with on a case-by-case basis. The box plot of the wading site reference data shows differences and similarities between ecoregions and the transition from lower scores in the Huron/Erie Lake Plain ecoregion to higher scores in the other four ecoregions. A similar stepwise procedure was used to calibrate the Invertebrate Community Index for macroinvertebrates (DeShon, 1995; Ohio Environmental Protection Agency, 1987b) and derive the quantitative biological criteria for the Modified Index of Well-Being for fish assemblages (Ohio Environmental Protection Agency, 1987b; Yoder and Rankin, 1995a).

TABLE 2   Calibrated Index of Biotic Integrity Modified for Ohio, Wading Sites

	Score		
Metric	5	3	1
1. Number of native species	Varies × Drainage Area		
2. Number of darter species	Varies × Drainage Area		
3. Number of sunfish species	>3	2–3	<2
4. Number of sucker species	Varies × Drainage Area		
5. Intolerant species			
>100 sq. mi. watershed	>5	3–5	<3
<100 sq. mi. watershed	Varies × Drainage Area		
6. Percent tolerant species	Varies × Drainage Area		
7. Percent omnivores	<19	19–34	>34
8. Percent insectivores			
<30 sq. mi. watershed	Varies × Drainage Area		
>30 sq. mi. watershed	>55	26–55	<26
9. Percent top carnivores	>5	1–5	<1
10. Abundance	>750	200–750	<200
11. Percent simple lithophils	Varies × Drainage Area		
12. Percent DELT anomalies	>1.3	0.5–1.3	<0.5

NOTE: DELT = animals with deformities (D), eroded (E) fins or barbels, lesions (L), or tumors (T).

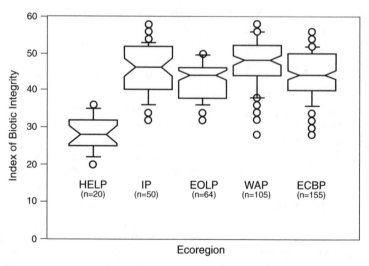

FIGURE 4   Notched box-and-whisker plot of reference-site results for the Index of Biotic Integrity from wading sites in each of the five ecoregions: Huron/Erie Lake Plain (HELP), Interior Plateau (IP), Eastern-Ontario Lake Plain (EOLP), Western Allegheny Plateau (WAP), and Eastern Corn Belt Plains (ECBP).

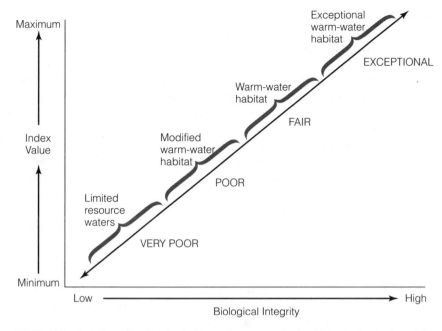

FIGURE 5   Relationship of biological integrity to quantitative biological criteria and the habitat classifications in the Ohio Water-Quality Standards. The corresponding relationship to narrative categories of biological community performance is also shown.

Figure 5 shows the relationships between the various habitat classifications, relative biological integrity, and the biological index values used to express the quantitative biological criteria. Narrative ratings of biological community performance are given opposite the habitat classifications. The highest biological index values coincide with the highest degree of biological integrity and the designation of exceptional warm-water habitat. Lower biological index values coincide with lesser degrees of biological integrity and poorer habitat classifications. As a matter of policy, only the criteria for warm-water habitat and exceptional warm-water habitat are considered consistent with the biological integrity goal of the Clean Water Act. However, states may designate poorer classifications such as modified warm-water habitat or limited resource waters if they can demonstrate that attempts to attain better biological integrity would cause substantial adverse socioeconomic impacts.

APPLICATIONS OF QUANTITATIVE BIOLOGICAL CRITERIA

Once biological criteria are derived and codified in state regulations, they are ready to use in water-quality management. Biological criteria need not be included in the water-quality standards to use them as an assessment tool, but in our experience, adopting them in water-quality standards significantly broadens and legitimizes their full use. Biological criteria are employed principally as an ambient monitoring and assessment tool through biological surveys. Biological and water-quality surveys, or biosurveys, are monitoring efforts on a water-body or watershed scale. They may range from a focus on a relatively simple setting with one or two small streams, one or two principal stressors, and a handful of sampling sites, to a much more complex effort that includes entire drainage basins, multiple and overlapping stressors, and dozens of sites. Each year, the Ohio Environmental Protection Agency conducts biosurveys in 10–15 different study areas with a total of 250–300 sampling sites.

The Ohio Environmental Protection Agency employs biological, chemical, and physical monitoring and assessment techniques in biosurveys to meet three major objectives:

• determine the extent to which the habitat classification-specific biological criteria are met;
• determine if habitat classification criteria assigned to each water body are appropriate and attainable; and
• determine if any changes in biological, chemical, or physical indicators have occurred since earlier measurements, particularly before and after the implementation of point-source pollution controls or best-management practices for nonpoint sources.

Identifying the causes of observed impairments requires the interpretation of multiple lines of evidence, including water chemistry data, sediment data, habitat data, effluent data, biomonitoring results, and land-use data (Yoder and Rankin, 1995b). The assignment of principal causes of impairment represents the association of impairments with stressor and exposure indicators. The principal reporting venue for this process is a biological and water-quality report for each watershed or subbasin. These reports include summaries of major findings and recommendations for revisions to water-quality standards, future monitoring needs, or other actions needed to resolve existing impairment. Although the principal focus of a biosurvey is to assess whether conditions meet biological criteria, they also address the status of the site for other uses, such as recreation and water supply. Such reports provide the foundation for aggregated assessments, including the Ohio Water Resource Inventory, the Ohio Nonpoint Source Assessment, and other technical bulletins.

Interpreting Results on the Basis of a Longitudinal Reach or Subbasin

It is often useful to plot results as a function of sampling location with major sources of potential impact and the applicable quantitative biological criteria indicated. Figure 6 shows fish and macroinvertebrate community data for the Scioto River during 1980 and 1991. This type of analysis is critical for demonstrating changes through time. Figure 6 represents the results of bioassessments in a 40-

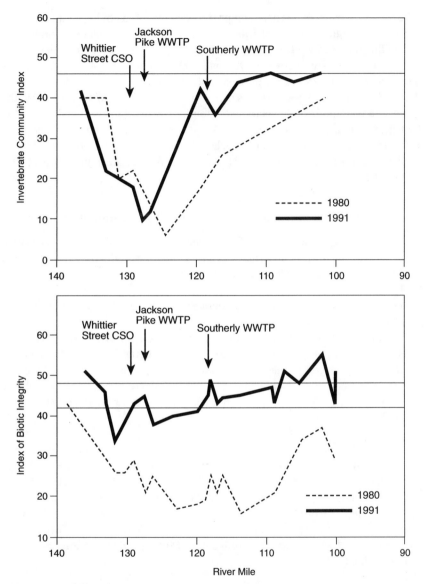

FIGURE 6   Values of the Invertebrate Community Index and the Index of Biotic Integrity for the Scioto River between Columbus and Circleville, Ohio, based on electrofishing samples collected July–October 1980 and 1991.   NOTE: The locations of the Whittier Street combined sewer overflow (CSO) and the Jackson Pike and Southerly waste-water treatment plants (WWTP) are indicated.   Horizontal lines indicate the criteria for exceptional warm-water habitat and warm-water habitat designations for the two indices.   The exceptional warm-water habitat criterion is 46 for the Invertebrate Community Index and 48 for the Index of Biotic Integrity.   The warm-water habitat criterion is 36 for the Invertebrate Community Index and 42 for the Index of Biotic Integrity.

mile river segment that has been sampled repeatedly. The obvious improvements exhibited by both the Index of Biotic Integrity and the Invertebrate Community Index illustrate the benefits of improved municipal waste-water treatment in the Columbus, Ohio, area between 1980 and 1991. Improvements designed to reduce water pollution were put in place at the major waste-water treatment plants during this interval. These changes can also be displayed in tabular form, but the extent and magnitude of the incremental improvements along each index axis can be better demonstrated graphically. Despite improvements between 1980 and 1991, Figure 6 shows that some reaches do not meet their respective criteria, particularly those close to combined sewer overflows and areas subjected to habitat modification.

Determination of Attainment of Habitat Classification Biocriteria

Once metric values have been calculated, each sampling site is classified according to the following criteria (Ohio Environmental Protection Agency, 1987b; Yoder, 1991b; Yoder and Rankin, 1995a):

• Full attainment—All of the applicable biological indices meet their respective criteria.
• Partial attainment—Either the fish or the macroinvertebrate assemblage fails to satisfy its criteria, but with biological index scores that are "fair" or better, whereas the other group (macroinvertebrates or fish) does satisfy its criteria.
• Nonattainment—Neither the fish nor the macroinvertebrate assemblage meets its biocriteria, or one of the assemblages reflects a narrative rating of "poor" or "very poor."

The analysis results provide the formal basis for determining whether biological criteria have been attained and serve as the starting point for all other uses of the data, including reporting (e.g., basin reports, Clean Water Act 305[b] report) and assessment (e.g., Water Quality Permit Support Documents).

The results of biosurveys can also be summarized in terms of Biological Integrity Equivalents (BIEs), a product of results from the Index of Biotic Integrity (IBI), the Modified Index of Well-Being (MIwb), and the Invertebrate Community Index (ICI):

$$BIE = \frac{(IBI + [MIwb \times 5] + ICI)}{180} \times 100.$$

Because the maximum values for IBI and ICI are 60, but the maximum theoretical value of MIwb is 12, MIwb is multiplied by 5 so that equal weight is given to each index. Their sum is then divided by 180, and the result is multiplied by 100 to produce scores that range from 0 to 100. The BIE score, being the composite of the three indices that comprise the biological criteria, reflects the degree to which a particular sample achieves the ideal of biological integrity. Figure 7 provides BIE results for the Kokosing and Hocking rivers in central Ohio. These rivers represent extremes in terms of impairment, water quality, aquatic community potential, and habitat classifications.

The Kokosing River figure shows longitudinal (i.e., upstream to downstream) changes in habitat classification, application of the quantitative biological criteria, the position of significant influences on water quality, and a change in ecoregion. One result of the 1987 survey was a proposal to change the original warm-water habitat designation to exceptional warm-water habitat for river miles 30 to 0 because the sampling results demonstrated the full attainment of the exceptional warm-water habitat biological criteria for both fish and macroinvertebrates in this area. The existing warm-water habitat designation originally was made in 1985 without the benefit of site-specific biological data.

The Hocking River presents a stark contrast to the Kokosing. The extensive nonattainment observed in 1982 was largely due to point-source discharges, combined sewers, urbanization, and habitat impacts. Improvements in municipal waste-water treatment and industrial pretreatment were primarily responsible for the improved conditions observed in 1990. This example shows how biological criteria can serve as a feedback tool for determining the success of pollution control programs.

Statewide Reporting and Assessment Applications

Biological data and biological criteria are the principal arbiters of habitat classification attainment status for the biennial Ohio Water Resource Inventory (Clean Water Act 305[b] report). Perhaps the question we are most frequently asked is, Is water quality improving? We have developed a standardized approach to biological monitoring and a long-term database that enables us to answer this and other questions. Figures 8 and 9 and Table 3 show data that enable us to track changes and document improvements in water quality. Although the Ohio Environmental Protection Agency database was not collected under a statistically random design and is spatially biased, the large number of sites (>4,500 locations) and thorough coverage of streams and rivers with drainage areas greater than 100 mi^2 (71 percent coverage statewide) make valid statewide comparisons possible. (The Ohio Environmental Protection Agency database represents a sampling site density more than 10 times that of the EPA Environmental Monitoring and Assessment Program and the U.S. Geological Survey National Water Quality Assessment monitoring designs.)

To answer the question of whether water quality is improving, we compared the direction and magnitude of change in biological index scores at sites for which we have multiple years of data. In this analysis, trends represent the difference between the earliest and latest results (most of which are approximately 10 years

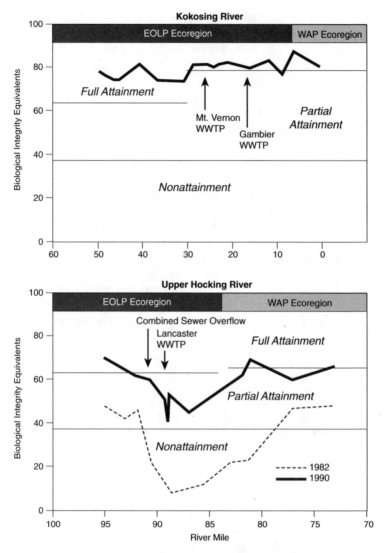

FIGURE 7   Longitudinal profile of Biological Integrity Equivalents (BIEs) for the Kokosing River in 1987 and the upper Hocking River in 1982 and 1990. NOTE: The BIE scores that generally coincide with full, partial, and nonattainment of the applicable habitat criteria are indicated. The Kokosing River is designated as exceptional warm-water habitat from mile 0 to mile 30 and warm-water habitat from mile 30 to mile 60. The section of the Hocking River represented here is designated as warm-water habitat. Locations of a combined sewer overflow and waste-water treatment plants (WWTP) are indicated in the figure. Shading at the top of each panel identifies river segments in the Eastern-Ontario Lake Plain (EOLP) and the Western Allegheny Plateau (WAP) ecoregions, respectively.

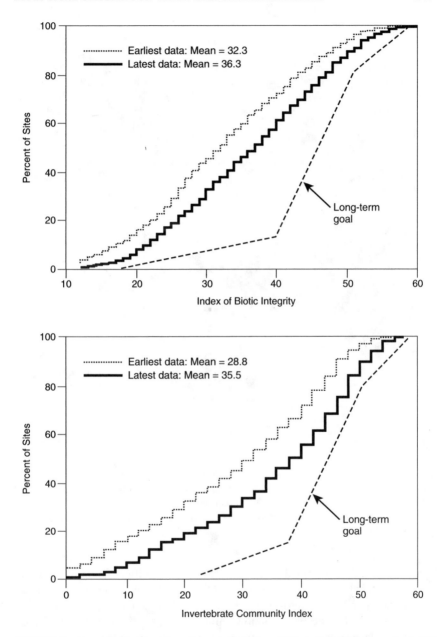

FIGURE 8 Cumulative frequency diagram for Index of Biotic Integrity scores for 1,160 Ohio sites and Invertebrate Community Index scores for 854 Ohio sites measured before and after 1988. NOTE: Measurements from before 1988 are represented by the line labeled "Earliest Data." Measurements from after 1988 are represented by the line labeled "Latest Data."

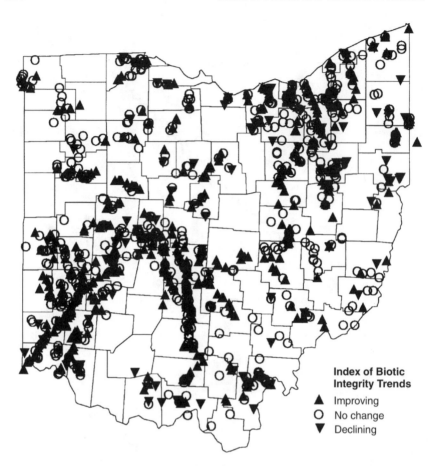

**Index of Biotic
Integrity Trends**

▲ Improving
O No change
▼ Declining

FIGURE 9   Map of sampling sites on Ohio rivers and streams depicting changes in the
Index of Biotic Integrity (IBI) at 1,160 sites and Invertebrate Community Index (ICI) at
527 sites. NOTE: Symbols indicate the directions of changes in the two indicators at the
various sites based upon a comparison of measurements from the earliest and latest sampling years before and after 1988.

apart). The analysis included 1,160 sites for the Index of Biotic Integrity, 845 sites
for the Modified Index of Well-Being, and 527 sites for the Invertebrate Community Index. Significant improvements have been observed for each index (Table 3;
Figures 8 and 9). The Invertebrate Community Index showed both the largest
increase and the largest shift in the frequency of sites entering the good and exceptional performance ranges (i.e., scores meeting the warm-water habitat and exceptional warm-water habitat criteria), but the fewest sites exiting the poor and very
poor ranges (scores <14). In contrast, the fish community Index of Biotic Integrity

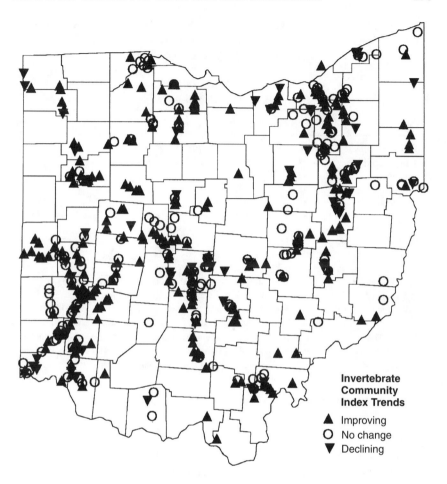

Invertebrate Community Index Trends

▲ Improving
○ No change
▼ Declining

had the greatest number of sites exiting the poor and very poor ranges (scores >26–28) but the fewest sites entering the good and exceptional ranges.

Considerable improvements have been documented in Ohio rivers and streams, as reflected by the biological criteria. The predominant pattern begins with the recovery of the macroinvertebrate community (as measured by the Invertebrate Community Index), followed later by improvements in fish abundance and biomass (as indicated by the Modified Index of Well-Being), and then finally structural and functional improvements (as measured by the Index of Biotic Integrity).

TABLE 3   Summary of Results for Ohio Stream and River Sites with at
Least 2 Years of Biological Data Collected between 1979 and 1994, at
Least Once Before (earliest) and Once After (latest) 1988

Category	Index of Biotic Integrity		Modified Index of Well-Being		Invertebrate Community Index	
	Earliest	Latest	Earliest	Latest	Earliest	Latest
10th percentile	16	20	3.8	4.8	6	14
25th percentile	24	28	5.7	6.6	16	26
Median	32	36	7.4	8.1	32	38
75th percentile	42	45	8.6	9.2	42	46
90th percentile	48	50	9.3	9.9	46	52
Mean	32.2	36.2	6.91	7.72	28.9	35.5
Degrees of freedom	1159		844		527	
t value	16.89		14.34		12.53	
Mean difference	3.95		0.80		6.65	
t-test P value	P <0.0001		P <0.0001		P <0.0001	
Wilcoxon (Z)	24.40		13.50		11.55	
Wilcoxon text P value	P <0.0001		P <0.0001		P <0.0001	

NOTE: Data pairs show descriptive statistics for earliest and latest measurements of the Index
of Biotic Integrity, the Modified Index of Well-Being, and the Invertebrate Community Index at
the various sites. Paired t-test and Wilcoxon's Z-test statistics compare the means of the earliest
and latest measurements of each of the indices.

SOURCE: Ohio Environmental Protection Agency (1996).

These substantial improvements notwithstanding, a significant proportion of
Ohio's rivers and streams remain too polluted and/or physically degraded to meet
their biological criteria. Figure 10 illustrates the aggregate changes in impairment that took place between 1988 and 1996 and projects the trend through the
year 2002. Figure 10 makes it clear that the proportion of impairment associated
primarily with point-source discharges is declining at a more rapid rate than impairment associated with nonpoint sources. Nonpoint sources include habitat
modifications, nutrient enrichment, and sedimentation. The state will need to pay
more attention to nonpoint sources and watershed-level effects if it is to reach
milestones such as the Ohio 2000 goal of 75 percent of full attainment. This will
require a significant restructuring of state water-quality management programs
that are presently heavily oriented toward point sources of pollutants.

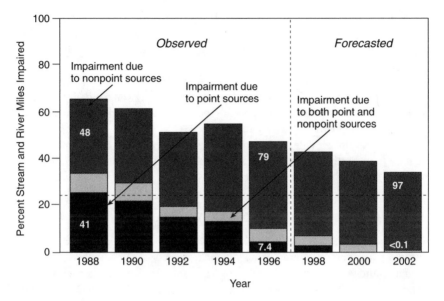

FIGURE 10   Observed and forecasted reductions, by percent, in the proportion of river and stream miles failing to attain their habitat classification biological criteria. NOTE: Forecasts from 1997 to 2002 are based on the observed rate of restoration during the period 1988–1996. The dashed horizontal line represents the Ohio 2000 goal.

OVERALL IMPACT OF BIOLOGICAL CRITERIA

The use of biological criteria has proved useful to the Ohio Environmental Protection Agency for several reasons:

• The results use direct measurements of ecological condition, rather than surrogate or symptomatic measures. This helps focus management programs on actual environmental results rather than administrative goals only (i.e., number of permits issued, grant dollars awarded, etc.).
• The resulting knowledge of aquatic community conditions can more efficiently guide management and regulatory activities that might otherwise be forced to rely on prescriptive approaches to information gathering (e.g., effluent characterization, major discharger lists, 303[d] and 304[l] lists).
• The results provide objective measurements with which to assign appropriate habitat classifications to individual rivers and streams.
• The results provide a means to assess the applicability and effectiveness of the antidegradation policy in the Ohio water-quality standards (i.e., extending antidegradation concerns to nonpoint sources and habitat influences, defining high-quality waters).

• Chemical and narrative water-quality criteria and standards can be more appropriately applied and take into account the integrated dynamics of the receiving waters when relevant biological assessment information is available (Yoder, 1991a,b).

• The biological results provide a legal basis for enforcement against entities discharging chemicals for which there are no existing water-quality standards or effluent guidelines (or at least provide the impetus to designate new chemical criteria or whole-effluent toxicity limits).

• The results provide a basis for regulating nonchemical environmental degradation (e.g., certifications of dredging permits [Section 401 water-quality certifications], non-point-source management).

REMAINING CHALLENGES

We have demonstrated how biological criteria can be developed and used within a state water-resource-management framework. Nonetheless, some important challenges remain. The cumulative costs associated with environmental mandates, many of which consist of prescription-based regulations, have recently come into question. Both the regulated community and the public desire evidence of real-world results in return for the expenditures made necessary by federally and state-mandated requirements. Biological criteria seem particularly well suited to address these concerns because the underlying science and theory is robust (Karr, 1991), and biocriteria directly assess the biological condition of aquatic habitats.

Although no single environmental indicator can do it all, biological criteria have a major role to play. A lack of information from or an overreliance on any single class of indicators can result in environmental regulation that is inaccurate and either under- or overprotective of the resource. Accounting for cost is not only a matter of dollars spent but also of program effectiveness. A credible and genuinely cost-effective approach to water-quality management should include an appropriate mix of chemical, physical, and biological measures, each in their respective roles as stressor, exposure, and response indicators. The public must come to see comprehensive monitoring designs using such cost-effective indicators as a part of the cost of doing business, perhaps at the expense of other programs when new evidence suggests that the resources allocated are disproportionate to the magnitude of the present problems (e.g., point versus nonpoint sources).

Based on our experience over the past 17 years, it is evident that including biological criteria in a state's monitoring and assessment effort has multiple benefits: It can foster a more complete integration of important ecological concepts, better focus water resource policy and management, and enhance strategic planning. Some specific examples include:

• Watershed approaches to monitoring, assessment, and management—The monitoring and assessment design inherent to biological criteria is fundamentally oriented to yield information on watersheds.

• Integrated point, nonpoint, and habitat assessment and management—Biological criteria integrate the effects of all stressors over time and space. The attendant use of chemical, toxicological, and physical tools can connect probable causes to observed impairments. This should provide a firm setting for the collaborative use of the same information for the management and regulation of both point and nonpoint sources (including habitat), two realms that have thus far been treated independently.

• Cumulative effects—Biological communities inhabit the receiving waters all of the time and reflect the integrative, cumulative effect of various stressors.

• Biodiversity issues—The basic biological data provide information about species, populations, and communities of concern.

• Interdisciplinary focus—Because biosurvey monitoring and assessment design is inherently interdisciplinary, the biological criteria approach provides the opportunity to bring ecologists, toxicologists, engineers, and other professionals together in planning and conducting assessments, interpreting results, and using information in strategic planning and management actions.

Biological criteria are an emerging and increasingly important issue for the EPA, the states, and the regulated community, and their use is growing nationwide. However, much remains to be done, particularly in the area of national and regional leadership. Technical guidance and expertise is needed to ensure a nationally consistent and credible approach and to resolve outstanding technical concerns (Yoder and Rankin, 1995a). Outstanding policy issues, such as the EPA's policy of independent applicability, need to be resolved in a manner that will encourage states to participate. In an era of declining government resources, we must develop ways to do more biological monitoring to support the biocriteria approach. Based on our experience in Ohio, the staffing of state biological assessment programs should include a minimum of one work year equivalent for every 1,200 miles of perennial streams and rivers. This estimate may vary by region and should incorporate lake-surface area in states with a substantial number of lakes (Yoder and Rankin, 1995a). The EPA must consider the potential for bioassessments and biocriteria to modify the present capital- and resource-intensive system of tracking environmental compliance on the basis of specific pollutants. The biological criteria approach should prove a more cost-effective way of managing the nation's water-quality programs.

ACKNOWLEDGMENTS

This paper would not have been possible without the many years of fieldwork, laboratory analysis, and data assessment and interpretation performed by members (past and present) of the Ohio Environmental Protection Agency, Ecological Assessment Section. Several staff members contributed extensively to the development of the biological assessment program, including Dave Altfater, Randy Sanders, Marc Smith, and Roger Thoma (fish methods, MIwb, and IBI metrics development) and Mike Bolton, Jeff DeShon, Jack Freda, Marty Knapp, and Chuck McKnight (macroinvertebrate methods and ICI metrics development). None of this effort would have been possible without the excellent data management and processing skills contributed by Dennis Mishne. Other staff (past and present) who made important contributions include Paul Albeit, Ray Beaumier, Chuck Boucher, Bernie Counts, Beth Lenoble, and Paul Vandermeer. Dan Dudley and Jim Luey contributed extensively to the early development and review of the then-emerging concepts of biological integrity, ecoregions, reference sites, and biological assessment in general. Charlie Staudt provided many hours of support in the development of the computer programs used for data analysis. Finally, Gary Martin and the late Pat Abrams provided solid management support for the concept of biological criteria and biological assessment at the Ohio Environmental Protection Agency.

REFERENCES

Angermier, P. L., and J. R. Karr. 1986. Applying an index of biotic integrity based on stream-fish communities: Considerations in sampling and interpretation. North American Journal of Fisheries Management 6:418–427.

Angermier, P. L., and J. E. Williams. 1993. Conservation of imperiled species and reauthorization of the endangered species act of 1973. North American Journal of Fisheries Management 18(7):34–38.

Ballentine, R. K., and L. J. Guarria. 1975. Paper published in The Integrity of Water: Proceedings of a Symposium, March 10–12, 1975, Washington, D.C. U.S. Environmental Protection Agency, Office of Water and Hazardous Materials. Washington, D.C.: U.S. Government Printing Office.

Barbour, M. T., J. Gerritsen, G. E. Griffin, R. Frydenborg, E. McCarron, J. S. White, and M. L. Bastian. 1996. A framework for biological criteria for Florida streams using benthic macroinvertebrates. Journal of the North American Benthological Society 15(2):185–211.

Benke, A. C. 1990. A perspective on America's vanishing streams. Journal of the North American Benthological Society 9(1):77–88.

Bennet, M. R., J. W. Kleene, and V. O. Shanholtz. 1993. Total maximum daily load nonpoint source allocation pilot project. File Report. Blacksburg, Va.: Virginia Department of Agricultural Engineering.

Bramblett, R. G., and K. D. Fausch. 1991. Variable fish communities and the index of biotic integrity in a western Great Plains river. Transactions of the American Fisheries Society 120:752.

Davies, S. P., L. Tsomides, D. L. Courtemanch, and F. Drummond. 1993. State of Maine Biological Monitoring and Biocriteria Development Program Summary. Augusta: Maine Department of Environmental Protection.

Davis, W. S., and A. Lubin. 1989. Statistical validation of Ohio EPA's invertebrate community index. Pp. 23–32 in Proceedings of the 1989 Midwest Pollution Biology Meeting, Chicago, Ill., W. S. Davis and T. P. Simon, eds. EPA 905/9-89/007. Washington, D.C.: U.S. Environmental Protection Agency.

DeShon, J. D. 1995. Development and application of the invertebrate community index (ICI). Pp. 217–243 in Biological Assessment and Criteria: Tools for Risk-Based Planning and Decision Making, W. S. Davis and T. Simon, eds. Boca Raton, Fla.: Lewis Publishers.

Fore, L. S., J. R. Karr, and L. L. Conquest. 1993. Statistical properties of an index of biotic integrity used to evaluate water resources. Canadian Journal of Fisheries and Aquatic Sciences 51:1077–1087.

Gakstatter, J., J. R. Gammon, R. M. Hughes, L. S. Ischinger, M. Johnson, J. Karr, T. Murphy, T. M. Murray, and T. Stuart. 1981. A recommended approach for determining biological integrity in flowing waters. Corvallis, Ore.: U.S. Environmental Protection Agency.

Gammon, J. R. 1976. The fish populations of the middle 340 km of the Wabash River. Water Resources Research Center Technical Report 86. West Lafayette, Ind.: Purdue University.

Gammon, J. R., A. Spacie, J. L. Hamelink, and R. L. Kaesler. 1981. Role of electrofishing in assessing environmental quality of the Wabash River. Pp. 307–324 in Ecological Assessments of Effluent Impacts on Communities of Indigenous Aquatic Organisms, J. M. Bates and C. I. Weber, eds. ASTM STP 730. Philadelphia, Pa.: American Society for Testing and Materials.

Hughes, R. M., D. P. Larsen, and J. M. Omernik. 1986. Regional reference sites: A method for assessing stream pollution. Environmental Management 10:629.

Intergovernmental Task Force on Monitoring Water Quality. 1992. Ambient Water Quality Monitoring in the United States: First Year Review, Evaluation, and Recommendations. Washington, D.C.: Interagency Advisory Committee on Water Data.

Intergovernmental Task Force on Monitoring Water Quality. 1993. Ambient Water Quality Monitoring in the United States: Second Year Review, Evaluation, and Recommendations and Appendices. Washington, D.C.: Interagency Advisory Committee on Water Data.

Intergovernmental Task Force on Monitoring Water Quality. 1995. The Strategy for Improving Water-Quality Monitoring in the United States. Final report of the Intergovernmental Task Force on Monitoring Water Quality and Appendices. Washington, D.C.: Interagency Advisory Committee on Water Data.

Judy, R. D. Jr., P. N. Seely, T. M. Murray, S. C. Svirsky, M. R. Whitworth, and L. S. Ischinger. 1984. 1982 National Fisheries Survey, Vol. 1, Technical Report Initial Findings. FWS/OBS-84/06. Washington, D.C.: U.S. Fish and Wildlife Service.

Karr, J. R. 1981. Assessment of biotic integrity using fish communities. Fisheries 6(6):21–27.

Karr, J. R. 1991. Biological integrity: A long-neglected aspect of water resource management. Ecological Applications 1(1):66–84.

Karr, J. R., and D. R. Dudley. 1981. Ecological perspective on water quality goals. Environmental Management 5:55–68.

Karr, J. R., K. D. Fausch, P. L. Angermier, P. R. Yant, and I. J. Schlosser. 1986. Assessing biological integrity in running waters: A method and its rationale. Illinois Natural History Survey Special Publication 5. Champaign, Ill.: Illinois Natural History Survey.

Kerans, B. L., and J. R. Karr. 1992. An Evaluation of Invertebrate Attributes and a Benthic Index of Biotic Integrity for Tennessee Valley Rivers. Proceedings of the 1991 Midwest Pollution Biology Conference. EPA 905/R-92/003. Chicago, Ill.: U.S. Environmental Protection Agency.

Leonard, P. M., and D. J. Orth. 1986. Application and testing of an index of biotic integrity in small, coolwater streams. Transactions of the American Fisheries Society 115:401–414.

Lyons, J. 1992. Using the index of biotic integrity (IBI) to measure environmental quality in warmwater streams of Wisconsin. General Technical Report NC-149. St. Paul, Minn.: U.S. Department of Agriculture, Forest Service, North Central Forest Experimental Station.

Lyons, J., and L. Wang. 1996. Development and validation of an index of biotic integrity for coldwater streams in Wisconsin. North American Journal of Fisheries Management 16:241–256.

Miller, D. L., P. M. Leonard, R. M. Hughes, J. R. Karr, P. B. Moyle, L. H. Schrader, B. A. Thompson, R. A. Daniel, K. D. Fausch, G. A. Fitzhugh, J. R. Gammon, D. B. Halliwell, P. L. Angermeier, and D. J. Orth. 1988. Regional applications of an index of biotic integrity for use in water resource management. Fisheries 13:12–20.

Oberdorff, T., and R. M. Hughes. 1992. Modification of an index of biotic integrity based on fish assemblages to characterize rivers of the Seine-Normandie basin, France. Hydrobiologia 228:116–132.

Ohio Environmental Protection Agency (OEPA). 1987a. Biological Criteria for the Protection of Aquatic Life, Vol. 1, The Role of Biological Data in Water Quality Assessment. Columbus, Ohio: OEPA, Division of Water Quality Monitoring and Assessment, Surface Water Section.

Ohio Environmental Protection Agency (OEPA). 1987b. Biological Criteria for the Protection of Aquatic Life, Vol. 2, Users Manual for Biological Field Assessment of Ohio Surface Waters. Columbus, Ohio: OEPA, Division of Water Quality Monitoring and Assessment, Surface Water Section.

Ohio Environmental Protection Agency (OEPA). 1988. Ohio Water Quality Inventory—1988 305(b) Report, Vol. 1 and Executive Summary, E. T. Rankin, C. O. Yoder, and D. A. Mishne, eds. Columbus, Ohio: OEPA, Division of Water Quality Monitoring and Assessment.

Ohio Environmental Protection Agency (OEPA). 1989a. Biological Criteria for the Protection of Aquatic Life, Vol. 3, Standardized Biological Field Sampling and Laboratory Methods for Assessing Fish and Macroinvertebrate Communities. Columbus, Ohio: OEPA, Division of Water Quality Monitoring and Assessment.

Ohio Environmental Protection Agency (OEPA). 1989b. Addendum to Biological Criteria for the Protection of Aquatic Life, Vol. 2, Users Manual for Biological Field Assessment of Ohio Surface Waters. Columbus, Ohio: OEPA, Division of Water Quality Planning and Assessment, Surface Water Section.

Ohio Environmental Protection Agency (OEPA). 1990. Ohio Water Resource Inventory, Vol. 1, Summary, Status, and Trends, E. T. Rankin, C. O. Yoder, and D. A. Mishne, eds. Columbus, Ohio: OEPA, Division of Water Quality Planning and Assessment.

Ohio Environmental Protection Agency (OEPA). 1992. Biological and Habitat Investigation of Greater Cincinnati Area Streams: The Impacts of Interceptor Sewer Line Construction and Maintenance, Hamilton and Clermont Counties, Ohio. OEPA Technical Report EAS/1992-5-1. Columbus, Ohio: OEPA.

Ohio Environmental Protection Agency (OEPA). 1995. Ohio Water Resource Inventory, Vol. 1, Summary, Status, and Trends, E. T. Rankin, C. O. Yoder, and D. A. Mishne, eds. Ohio EPA Technical Bulletin MAS/1994-7-2-I. Columbus, Ohio: OEPA, Division of Surface Water.

Ohio Environmental Protection Agency (OEPA). 1996. Ohio Water Resource Inventory, Vol. 1, Summary, Status, and Trends, E. T. Rankin, C. O. Yoder, and D. A. Mishne, eds. Columbus, Ohio: OEPA.

Plafkin, J. L., M. T. Barbour, K. D. Porter, S. K. Gross, and R. M. Hughes. 1989. Rapid Bioassessment Protocols for Use in Rivers and Streams: Benthic Macroinvertebrates and Fish. EPA/444/4-89-001. Washington, D.C.: U.S. Environmental Protection Agency.

Rankin, E. T., and C. O. Yoder. 1990a. A Comparison of Aquatic Life Use Impairment Detection and Its Causes between an Integrated, Biosurvey-Based Environmental Assessment and Its Water Column Chemistry Subcomponent, Vol. 1, Appendix 1. Ohio Water Resource Inventory. Columbus, Ohio: Ohio Environmental Protection Agency, Division of Water Quality Planning and Assessment.

Rankin, E. T., and C. O. Yoder. 1990b. The nature of sampling variability in the index of biotic integrity (IBI) in Ohio streams. Pp. 9–18 in Proceedings of the 1990 Midwest Pollution Control Biologists Conference, W. S. Davis, ed. EPA-905-9-90/005. Chicago, Ill.: U.S. Environmental Protection Agency, Region V, Environmental Sciences Division.

Steedman, R. J. 1988. Modification and assessment of an index of biotic integrity to quantify stream quality in southern Ontario. Canadian Journal of Fisheries and Aquatic Sciences 45:492–501.

Stevens, J. C., and S. W. Szczytko. 1990. The use and variability of the biotic index to monitor changes in an effluent stream following wastewater treatment plant upgrades. Pp. 33–46 in Proceedings of the 1990 Midwest Pollution Biology Meeting, W. S. Davis, ed. EPA-905-9-90/ 005. Chicago, Ill.: U.S. Environmental Protection Agency.

Suter, G. W., II. 1993. A critique of ecosystem health concepts and indexes. Environmental Toxicology and Chemistry 12:1533–1539.

United States Environmental Protection Agency, Office of Water Regulations and Standards. 1990. Biological Criteria, National Program Guidance for Surface Waters. EPA-440/5-90-004. Washington, D.C.: U.S. Environmental Protection Agency.

United States Environmental Protection Agency, Office of Research and Development, Environmental Research Laboratory. 1991. Environmental Monitoring and Assessment Program. EMAP— Surface Waters Monitoring and Research Strategy—Fiscal Year 1991. EPA/600/3-91/022. Corvallis, Ore.: U.S. Environmental Protection Agency.

United States Environmental Protection Agency, Office of Water. 1994. National Water Quality Inventory: 1992 Report to Congress. EPA 841-R-94-001. Washington, D.C.: U.S. Environmental Protection Agency.

Yoder, C. O. 1989. The development and use of biological criteria for Ohio rivers and streams. In Water Quality Standards for the 21st Century. G. H. Flock, ed. Washington, D.C.: Office of Water, U.S. Environmental Protection Agency.

Yoder, C. O. 1991a. Answering some concerns about biological criteria based on experiences in Ohio. In Water Quality Standards for the 21st century, G. H. Flock, ed. Washington, D.C.: U.S. Office of Water, Environmental Protection Agency.

Yoder, C. O. 1991b. The integrated biosurvey as a tool for evaluation of aquatic life use attainment and impairment in Ohio surface waters. In Biological Criteria: Research and Regulation. EPA-440/5-91-005:110. Washington, D.C.: Office of Water, U.S. Environmental Protection Agency.

Yoder, C. O., and E. T. Rankin. 1995a. Biological criteria program development and implementation in Ohio. Pp. 109–144 in Biological Assessment and Criteria: Tools for Water Resource Planning and Decision Making, W. Davis and T. Simon, eds. Boca Raton, Fla.: Lewis Publishers.

Yoder, C. O., and E. T. Rankin. 1995b. Biological response signatures and the area of degradation value: New tools for interpreting multimetric data. Pp. 263–286 in Biological Assessment and Criteria: Tools for Water Resource Planning and Decision Making, W. Davis and T. Simon, eds. Boca Raton, Fla.: Lewis Publishers.

Measures of Environmental Performance and
Ecosystem Condition. 1999. Pp. 260–283.
Washington, DC: National Academy Press.

TVA's Approach to Ecological Health Assessment in Streams and Reservoirs

NEIL E. CARRIKER

The Tennessee Valley Authority (TVA) started a monitoring program in 1986 to evaluate the major tributaries of the Tennessee River at fixed locations. In 1990, a parallel reservoir monitoring program was begun. The combined stream and reservoir monitoring efforts consolidated several activities to form an integrated program that is part of TVA's comprehensive Clean Water Initiative.

The program's objectives are to provide information on the health or integrity of the aquatic ecosystem in major Tennessee River tributaries and reservoirs and to provide screening-level information to describe how well these water bodies meet the goals of fishability and swimmability set forth in the Clean Water Act. The TVA also carries out additional periodic monitoring of toxic contaminant levels in fish and bacteriological sampling at recreation areas to assess whether people can safely fish and swim in the Tennessee Valley waters. The ecological integrity of streams and reservoirs is evaluated as part of an activity called Vital Signs monitoring that is based on annual examinations of key physical, chemical, and biological indicators.

This paper focuses on how TVA develops ecological health ratings for reservoirs and streams. A technical report is published annually that further documents TVA's Vital Signs monitoring program and ecological health rating scheme and provides the most recent results from these efforts (Dycus and Meinert, 1991, 1992, 1993a,b, 1994).

VITAL SIGNS MONITORING

Premises

The Vital Signs monitoring study design is based on several fundamental premises:

- ecological health evaluations must be based on information on physical, chemical, and biological components of the ecosystem;
- monitoring must be sustained for several years to document the status of the river/reservoir system, determine its natural year-to-year variability, and track results of water-quality improvement efforts;
- monitoring must provide resource managers with current, useful information;
- monitoring program design must be dynamic and flexible rather than static and rigid and must allow resource managers to adopt better monitoring techniques as they develop to meet specific needs; and
- monitoring is not primarily intended to address specific cause-and-effect mechanisms. Although monitoring results may provide sufficient information to identify cause-and-effect relationships, addressing these mechanisms usually calls for shorter-term, more-detailed assessment programs.

With these fundamental premises in mind, TVA's challenge has been to develop a sustainable monitoring effort that collects the right kinds of data at a minimum number of locations and frequencies, yet still provides enough information to reliably characterize ecological health. The four main activities of the program focus on physical and chemical characteristics of water; acute toxicity and physical and chemical characteristics of sediment; benthic macroinvertebrate community sampling; and fish assemblage sampling. Under a complementary program, TVA also collects aquatic macrophyte community information to provide a more comprehensive evaluation of each reservoir's condition.

Monitoring Design

Sampling Locations

Three areas in each reservoir were selected for monitoring: the inflow area, which is generally riverine in nature; the transition zone or midreservoir area, where water velocity decreases due to increased cross-sectional area, suspended materials begin to settle, and algal productivity increases due to increased water clarity; and the forebay, the lacustrine area near the dam (Figure 1). Transition zone and forebay areas include overbanks (flood plains that are inundated when rivers are impounded).

Embayments constitute an important reservoir area that support a variety of uses. Previous studies have shown that ecological conditions in reservoir embayments are controlled mostly by the characteristics of the immediate watershed and embayment morphometry (Meinert et al., 1992). The main body of a reservoir usually has relatively little influence on embayment conditions because typically there is only minimal water exchange. But monitoring the ecological health of the hundreds of embayments in the TVA reservoir system is well beyond the scope of this program.

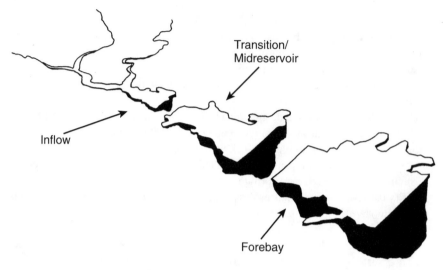

FIGURE 1 Key reservoir sampling areas.

Consequently, Vital Signs monitoring includes only four large embayments with drainage areas greater than 500 square miles and surface areas greater than 4,500 acres.

Locations for stream monitoring stations are chosen to sample as large a portion of each tributary watershed as possible. Stations generally are in un-impounded river reaches near the downstream end of each watershed.

Ecological Indicators

The selection of ecological indicators is tailored to the type of monitoring location. Physical, chemical, and biological indicators are used to provide information on the health of various habitats or ecological compartments. In reservoirs, the status of the open-water or pelagic area is represented by physical and chemical characteristics of water (including chlorophyll-a, a measure of phytoplankton abundance) and measurements of fish communities in midchannel. The shoreline or littoral area is evaluated by sampling the fish community. Two indicators provide information on the bottom or benthic compartment: quality of surface sediments in midchannel (chemical analysis of sediments and acute toxicity of pore water) and benthic macroinvertebrate community (10 samples collected across the full width of the reservoir at each station). In streams, all available habitats are sampled to fully characterize the station by measuring the same basic indicators used for reservoirs. For both reservoirs and streams, information from each indicator is evaluated separately, and results are then combined without weighting to arrive at an overall evaluation of ecological health.

Sampling Frequency

Sampling frequency reveals how each indicator varies over time. Physical and chemical indicators vary significantly in the short term; consequently, they are monitored monthly from spring to fall in reservoirs and every other month throughout the year in streams. Biological indicators better integrate long-term variations, and so reservoir and stream sites are sampled annually. Reservoir benthic macroinvertebrate sampling is conducted in early spring (February–April), and reservoir fish assemblage sampling is conducted in autumn (September–November). Benthic and fish community sampling in streams is conducted in late spring to early summer (May–June).

The net result of this monitoring design is that TVA collects data every year related to physical and chemical water-quality and biological conditions at sampling locations on 30 reservoirs (Figure 2) and 18 streams (Figure 3).

Aquatic Macrophytes

Aquatic macrophyte coverage is determined from large-scale (1 inch = 600 feet or 1 inch = 1,000 feet) color aerial photography taken during the late summer or early fall, which is the time of maximum submerged macrophyte growth. At approximately the same time as the overflight, boat surveys are conducted to determine macrophyte community structure at selected sites. Using Mylar overlays attached to photographic prints, aquatic macrophyte colonies are delineated and labeled according to species, and areas are measured using an electronic planimeter. Reservoirs flown for aerial photography usually include Kentucky, Wilson, Wheeler, Guntersville, Nickajack, Chickamauga, Tellico, South Holston, and lakes in the Beech River project. For reservoirs where aerial photography is not carried out, standard field surveys and historical information are used to estimate aquatic plant community structure and coverage. Submersed aquatic plant populations generally are rare in tributary reservoirs because of the wide fluctuations of water surface elevations associated with their operation for floodwater storage. Known populations have been extremely small, short-lived, and of little significance. A detailed summary of TVA's *Aquatic Plant Management Program* is available in a technical report (Burns et al., 1994) that is updated and published annually.

ECOLOGICAL HEALTH RATINGS

Approach

Because no official or universally accepted guidelines or criteria exist upon which to base an evaluation of ecological health, TVA developed the following methodology for rating the overall ecological health of each of the 30 TVA Vital Signs reservoirs and 18 Vital Signs stream stations. This method combines infor-

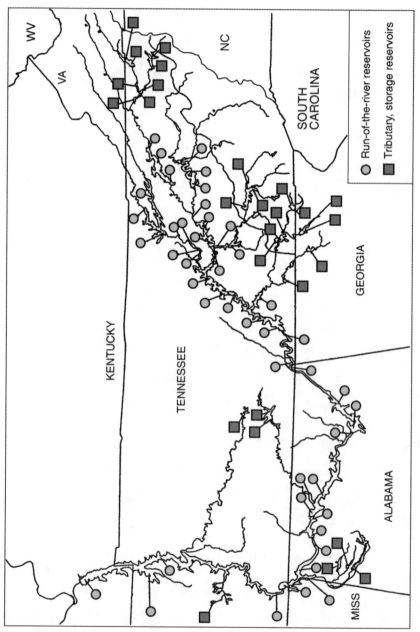

FIGURE 2 Reservoir Vital Signs monitoring locations, 1994.

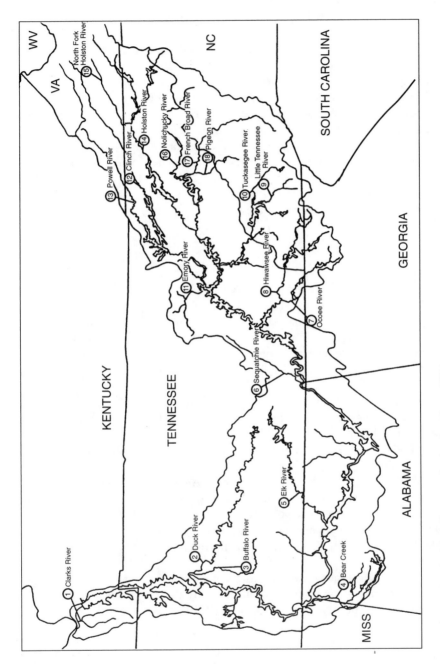

FIGURE 3 Stream Vital Signs monitoring locations, 1994.

mation on physical and chemical characteristics of water quality, biological community structure, and habitat. Of the many variables it collects, TVA has selected five indicators for evaluating reservoir health: dissolved oxygen, chlorophyll-*a*, sediment quality, benthic macroinvertebrates, and fish community. Stream evaluation is based on four aquatic ecosystem indicators: nutrient concentrations, sediment quality, benthic macroinvertebrates, and fish community.

Ecological health evaluations depend on the ability to discriminate between good and poor conditions for each indicator. This is more easily done for streams because they offer relatively unaltered reference sites that can be examined to define "good" conditions for each indicator. For example, various indices of biotic integrity for fish and benthic stream communities compare results at monitoring locations with conditions at reference sites (Karr et al., 1986; Kerans et al., 1992). But reservoirs are man-made alterations of natural streams; thus, no "reference reservoirs" exist for comparison. They require an alternative approach to reference conditions.

Overview

Scoring criteria for dissolved oxygen and chlorophyll-*a* in reservoirs are based on a conceptual model that TVA developed over several years from its experience in evaluating biological systems in reservoirs. The model for dissolved oxygen criteria for a reservoir is complicated by the combined effects of flow regulation and potential oxygen depletion in the hypolimnion (deep water). TVA's scoring criteria consider dissolved oxygen levels both in the water column and near the bottom of the reservoir. For chlorophyll-*a*, TVA's experience is that, below a threshold level, primary production is not sufficient to support an active, biologically healthy food chain. However, chlorophyll-*a* concentrations above a higher threshold result in undesirable eutrophic conditions.

For the benthic macroinvertebrate and fish community indicators, TVA based its scoring criteria on a statistical examination of multiple years of data from TVA reservoirs. All previously collected TVA reservoir data for a characteristic of a selected community (e.g., number of taxa, total abundance) are ranked and divided into good, fair, and poor groupings. The current year's results are compared with these groupings and scored accordingly. This approach is valid if the database is sufficiently large and covers the full spectrum of good-to-poor conditions.

The sediment-quality scoring criteria use a combination of sediment toxicity to test organisms and sediment chemical analyses for ammonia, heavy metals, pesticides, and polychlorinated biphenyls (PCBs).

Reservoir Scoring Criteria

Dissolved Oxygen

If only one indicator of reservoir health could be measured, dissolved oxygen (DO) would likely be the indicator of choice. Hutchinson (1975) states that a series of oxygen measurements probably provides more information about the nature of a lake than any other kind of data. The presence, absence, and levels of DO in a lake or reservoir both control and are controlled by many physical, chemical, and biological processes (e.g., photosynthesis, respiration, oxidation-reduction reactions, bacterial decomposition). DO measurements, coupled with observations of water clarity (Secchi depth), temperature, nutrients, and some basic hydrologic and morphometric information, yield substantial information about the ecological health of a reservoir.

Ideally, a reservoir has near-saturation concentrations of DO available to fish, insects, and zooplankton throughout the water column. This is usually the case during winter and spring when most reservoirs are well mixed. However, summer brings more sunlight, warmer water, and lower flows. This causes thermal stratification and increased biological activity; these combine to produce a greater biochemical demand for oxygen, particularly in the deeper portions of the reservoir. As a result, summer levels of DO often are low in the metalimnion and hypolimnion (intermediate and deepest regions). Hypolimnetic and metalimnetic oxygen depletion are common but undesirable occurrences in many reservoirs, especially storage impoundments. If DO concentrations are low enough, or low concentrations are sustained long enough, the health and diversity of the fish and benthic communities suffer. Sustained near-bottom anoxia also promotes release of ammonia, sulfide, and dissolved metals into the interstitial pore water and near-bottom waters. If this phenomenon persists long enough, these chemicals can cause chronic or acute toxicity to bottom-dwelling animals.

Historic information for reservoirs in the Tennessee Valley reveals that the burrowing mayfly (*Hexagenia* sp.) disappears from the benthic community at DO concentrations of 2 mg/l and below (Masters and McDonough, personal communication, 1993). Most fish species avoid areas with DO concentrations below 2 mg/l; fish growth and reproduction decrease at these levels, and many highly desirable species such as sauger and walleye simply cannot survive at such low levels of DO. Consequently, TVA considers 2 mg/l a critical level for evaluating ecological health and has incorporated it into the scheme for rating DO.

The rating scheme considers oxygen concentrations both in the water column (WC_{DO}) and near the bottom of the reservoir (B_{DO}). The DO rating at each sampling location (ranging from 1, "poor," to 5, "good") weights equally the average monthly summer water column and bottom water DO concentrations. Summer is defined as a 6-month period when maximum thermal stratification and maximum hypolimnetic anoxia are expected to occur (April through Septem-

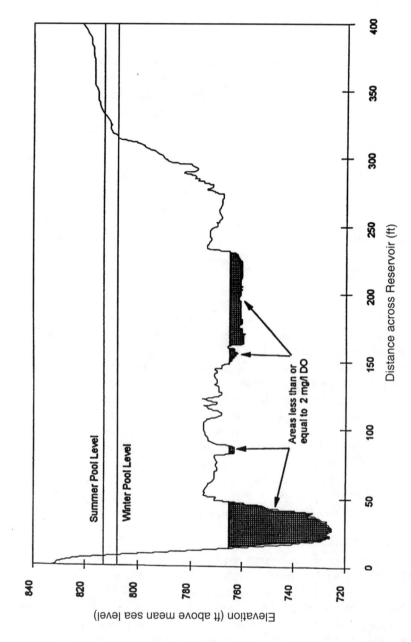

FIGURE 4 Cross-section of Tellico Reservoir forebay showing areas where summer DO concentrations averaged less than or equal to 2 mg/l.

TABLE 1   Cross-Sectional Area and Length Criteria (DO less than 2 mg/l) for WC_{DO} and B_{DO}

Average Cross-Sectional Area with DO less than 2 mg/l	WC_{DO} Rating
<5%	5 (good)
≥5% but ≤10%	3 (fair)
>10%	1 (poor)

Average Cross-Sectional Length with DO less than 2 mg/l	B_{DO} Rating
0%	5 (good)
>0 to 10%	4
>10 to 20%	3 (fair)
>20 to 30%	2
>30%	1 (poor)

ber for the run-of-the-river reservoirs and May through October for the tributary reservoirs).

WC_{DO} is a 6-month average of the percent of reservoir cross-sectional area at the sampling location that has a DO concentration less than 2.0 mg/l (Figure 4), and B_{DO} is a 6-month average of the percent of reservoir cross-sectional bottom length at the sampling location that has a DO concentration less than 2.0 mg/l. The overall DO rating is the average of the DO ratings of water column and the bottom. Several criteria are used to assign numerical ratings for WC_{DO} and B_{DO} (Table 1).

The average percent cross-sectional bottom length is computed based on the total cross-sectional bottom length at average minimum winter pool (water-level) elevation. In addition, if anoxic bottom conditions (0 mg/l) are observed, the B_{DO} rating for the site is lowered 1 unit, with a minimum rating of 1. In addition, because most state water-quality criteria for fish and aquatic life specify a minimum of 5.0 mg/l DO at a depth of 1.5 m (5 ft), the WC_{DO} rating also drops if any measured DO at that depth is below 5.0 mg/l (Table 2).

Chlorophyll-a

Chlorophyll-*a* is a simple and well-accepted measure for estimating algal biomass, algal productivity, and trophic condition (Carlson, 1977). Too little algal productivity in reservoirs indicates an inability to sustain a well-fed, growing, balanced, and healthy aquatic community, which eventually results in low standing stocks of fish. TVA's data suggest that a mean summer chlorophyll-*a* concentration less than 3 µg/l is a threshold below which ecological health is impaired. But too much primary productivity often results in dense algal blooms, poor water clarity, and the predominance of noxious blue-green algae, all indica-

TABLE 2   Relationship between Variable DO (at 1.5 m) and WC_{DO} Rating

Minimum DO at 1.5 m	WC_{DO} Rating Change
<5 mg/l	Decreased one unit (e.g., 5 to 4)
<4 mg/l	Decreased two units
<3 mg/l	Decreased three units
etc.	etc.

tors of poor ecological health. The large amounts of algal material produced under these conditions also deplete oxygen concentrations as the algae die and decompose. This can cause or aggravate problems of low DO in bottom waters. TVA results indicate that a mean summer chlorophyll-*a* concentration greater than 15 µg/l is a threshold above which these undesirable conditions are likely.

These threshold levels are incorporated into the chlorophyll-*a* ratings at each sampling location (Table 3). The average summer chlorophyll-*a* concentration of monthly photic zone samples collected from April through September (or October) is compared with these criteria and rated accordingly.

Sediment Quality

Contaminated bottom sediments can have direct adverse impacts on bottom fauna and can often be long-term sources of toxic substances in the aquatic environment. Wildlife and humans may be affected by these contaminants by ingestion or through direct contact. These effects may occur even though the water

TABLE 3   Threshold Chlorophyll-*a* Concentrations and Corresponding Ratings

Average Concentration[a]	Rating
<3 µg/l	3 (fair)[b]
3 to 10 µg/l	5 (good)
>10 to 15 µg/l	3 (fair)
>15 µg/l	1 (poor)

[a]If any single chlorophyll-*a* sample exceeds 30 mg/l, the value is not included in calculating the average, but the rating is decreased 1 unit (e.g., 5 to 4) for each sample that exceeded this value.

[b]If nutrients are sufficient (e.g., nitrate plus nitrite greater than 0.05 mg/l and total phosphorus greater than 0.01 mg/l) but chlorophyll-*a* concentrations are low (e.g., less than or equal to 2 µg/l), some other limiting or inhibiting factor such as toxicity is likely. When these conditions exist, chlorophyll-*a* is rated 2 (poor).

TABLE 4   Zooplankton Survival and Corresponding Sediment Toxicity
Ratings

S_{TOX} rating	Percent Survival of *Ceriodaphnia* and/or *Brachionus*
5 (good)	Survival not significantly different from control and greater than or equal to 80 percent for both species (i.e., no significant toxicity)
3 (fair)	Survival not significantly different from control, but less than 80 percent survival for either species
1 (poor)	Survival of either organism significantly less than control (i.e., significant toxicity)

above the sediments meets water-quality criteria. TVA's approach combines two assessment methods, one biological and one chemical, to evaluate reservoir sediment quality. TVA's scoring criterion is based on ratings for the toxicity of sediment pore water (S_{TOX}) to test organisms and the chemical analysis of sediment (S_{CHM}) for heavy metals, PCBs, organochlorine pesticides, and un-ionized ammonia. The final sediment-quality rating is the average of these two.

Sediment toxicity is evaluated using acute tests of survival of both rotifers (*Brachionus calyciflorus*) and daphnids (*Ceriodaphnia dubia*). A commercially available procedure (Rototox®) is used for the rotifer test and standard U.S. Environmental Protection Agency-approved procedures are used for the *Ceriodaphnia* test. These acute toxicity evaluations entail exposing these zooplankton to interstitial pore water from sediment. The survival rates of the organisms are based on the average survival in four replicates of five individuals each, compared with a control. If average survival is significantly reduced from the control, the sample is considered toxic (Table 4).

Sediment chemistry ratings (Table 5) are based on concentrations of heavy metals (Cd, Cr, Cu, Pb, Hg, Ni, and Zn) that exceed freshwater sediment guide-

TABLE 5   Sediment Chemistry Measurements and
Corresponding Ratings

S_{CHM} Rating	Sediment Chemistry
5 (good)	No measurements exceed guidelines
3 (fair)	One or two measurements exceed guidelines
1 (poor)	Three or more measurements exceed guidelines

lines (United States Environmental Protection Agency, 1977); detectable amounts of PCBs or pesticides; and concentrations of un-ionized ammonia in pore water above 200 mg NH_3/l.

Benthic Community

Six community characteristics (metrics), with scoring criteria specific to either run-of-the-river or storage reservoirs, are used to evaluate the ecological health of the benthic macroinvertebrate community:

• Taxa richness, or the number of different taxa present. More taxa indicate better conditions.
• Longed-lived species, or the number of taxa of *Corbicula, Hexagenia,* mussels, and snails present. Because these organisms are long-lived, their presence indicates conditions that allow long-term survival.
• EPT, or the number of different taxa present within the orders Ephemeroptera (mayflies), Plecoptera (stoneflies), and Trichoptera (caddisflies). Higher numbers indicate good water-quality conditions in streams. A similar use is incorporated here despite expected lower numbers in reservoirs than in streams.
• Proportion *Chironomidae,* or the percent of organisms present in the sample that are chironomids. A higher proportion indicates poor conditions.
• Proportion *Tubificidae,* or the percent of organisms present that are tubificids. A higher proportion indicates poor quality.
• Proportion as dominant taxa, or the percent of organisms present that are members of the dominant taxon. This metric is used as an evenness indicator. A large proportion composed of one or two taxa indicates poor conditions.

Results for 10 individual bottom samples collected at evenly distributed intervals along a transect at each station are compared with criteria for each metric. Scoring criteria for each metric have been developed for both run-of-the-river reservoirs and tributary reservoirs. Because of the substantial habitat differences among reservoir forebays, transition zones, and inflows, the scoring criteria also are stratified by area. Data handling differs somewhat among the metrics. Metric 1, taxa richness, is the total number of taxa for all 10 samples at each station. Metrics 2 and 3, long-lived species and EPT, are handled similarly. For metric 4, the proportion of chironomids in each sample is calculated. The proportions are then averaged, first for the station and then for the reservoir. An alternative that was considered and rejected was to sum the number of chironomids in all samples and divide by the total number of individuals for all samples. The approach selected gives equal weight to all samples regardless of sample size or sampling gear. This eliminates the bias introduced in the alternate approach, when one sample has an exceptionally high or low density. Metric 5, proportion *Tubificidae,* is calculated in the same way. Metric 6, proportion as dominant taxon, is calcu-

TABLE 6   Reservoir Area Metric Scores and Corresponding Ratings

Sum of Reservoir Benthic Community Metric Scores	Benthic Rating
6–10	1 (poor)
11–15	2
16–20	3 (fair)
21–25	4
26–30	5 (good)

lated similarly, using the proportion calculated for the dominant taxon in each sample, even if the dominant taxon differs among samples at a station. This allows more discretion to identify imbalances at a station than would developing an average for a single dominant taxon for all samples at the station.

The basis for evaluation criteria is the range of values found in the available database (all Vital Signs benthic monitoring data from 1991 to the present) for each metric. For each metric at each reservoir area (forebay, transition zone, and inflow) and reservoir type (run-of-the-river and tributary), the database values are divided into three groups using Ward's minimum variance analysis (SAS Institute, 1989). This procedure places observations into three homogeneous groups of approximately equal size. The groups are sorted and categorized as poor, fair, or good. Scoring criteria represent values between the highest and the lowest value in each group. The current year's results for each metric are compared with these criteria and assigned scores of 1 (poor), 3 (fair), or 5 (good) depending on which group they fall in. Scores are summed by reservoir area to yield an overall benthic rating for each location (Table 6).

Fish Assemblage

A Reservoir Fish Assemblage Index (RFAI) (Hickman et al., 1994) is used to rate fish assemblages as they relate to the overall ecological health of the reservoir. The RFAI is based on 12 metrics in 4 areas (Box 1) with scoring criteria specific to either run-of-the-river or tributary reservoirs. Scoring criteria also are specific for the type of sampling location within reservoirs—forebay, transition zone, or inflow; and for the type of sampling gear (i.e., electrofishing for littoral fish communities and gill netting for pelagic fish communities).

Each metric is assigned a score of 5, 3, or 1—representing good, fair, or poor conditions, respectively. Because of the distinct habitat differences among reservoirs and sampling locations (and the differences in the fish assemblages that they support), different scoring criteria are used according to reservoir type (run-of-the-river or tributary storage reservoirs), sampling location (forebay, transition, or inflow), and type of sampling gear (electrofishing or gill netting). There

---

**BOX 1   Reservoir Fish Assemblage Index—
Metric Characteristics**

Reservoir Fish Species Richness and Composition Metrics
- Total number of species—Greater numbers of species represent healthier aquatic ecosystems. As conditions degrade, species numbers decline.
- Number of piscivore species—Higher diversity of piscivores indicates a better quality environment.
- Number of sunfish species—Lepomid sunfish (excluding black bass, crappies, and rock bass) are basically insectivores; high diversity of this group indicates low levels of siltation and high sediment quality in littoral areas.
- Number of sucker species—Suckers also are insectivores, but they inhabit the pelagic and more riverine sections of reservoirs. This metric closely parallels the metric for lithophilic spawning species (below) and may be deleted from future RFAI calculations.
- Number of intolerant species—This group is made up of species that are particularly intolerant of habitat degradation. Higher numbers of intolerant species indicate better environmental quality.
- Percentage tolerant individuals (excluding young-of-year)—A high proportion of individuals tolerant of degraded conditions indicates poor environmental quality.
- Dominance by one species—Ecological quality is reduced if one species dominates the resident fish community.

---

is not yet enough information for inflow sampling locations on tributary reservoirs to establish criteria for fish community metrics at those sites.

The average of the sum of the electrofishing scores and the sum of the gill netting scores results in a RFAI for each station. The possible range of RFAI values is 12 (all metrics scored 1) to 60 (all metrics scored 5). This possible range is divided into five equal groupings to evaluate the overall health of the fish assemblage at each station (Table 7).

Discussions of the development of the RFAI and results of the fish evaluations for the 1991–1993 Vital Signs monitoring data are available in TVA technical reports (Brown et al, 1993; Hickman et al, 1994; Scott et al., 1992).

Overall Reservoir Health Determination

The methodology for evaluating overall ecological health combines the five indicators (DO, chlorophyll-*a*, sediment quality, benthic macroinvertebrates, and

Reservoir Fish Trophic Composition Metrics

- Percentage omnivores—Omnivores are less sensitive to environmental stresses due to their ability to vary their diets. As trophic links are disrupted by degraded conditions, specialist species such as insectivores decline, while opportunistic omnivorous species increase in relative abundance.

- Percentage insectivores—Due to the special dietary requirements of this group of species and the limitations of their food source in degraded environments, the proportion of insectivores increases with environmental quality.

Reservoir Fish Reproductive Composition Metric

- Number of lithophilic spawning species—Lithophilic broadcast spawners are sensitive to siltation. Numbers of lithophilic spawning species are higher in reservoirs with low rates of siltation.

Reservoir Fish Abundance and Fish Health Metrics

- Total catch per unit effort (number of individuals)—This metric assumes that high-quality fish assemblages support large numbers of individuals.

- Percentage with anomalies—Incidence of diseases, lesions, tumors, external parasites, deformities, blindness, and natural hybridization are noted for all fish measured, with higher incidence indicating poor environmental conditions.

fish assemblage) into a single numeric value. To arrive at this number, the first step is to sum the ratings for all indicators by station. Given the variations in both the number of indicators monitored at each station and the number of stations per reservoir, the ratings vary from 5 to 18 for the 30 reservoirs monitored in 1993.

TABLE 7   Reservoir Fish Assemblage Index and Corresponding Ratings

RFAI Score	Rating
12–21	1 (poor)
22–31	2
32–41	3 (fair)
42–51	4
52–60	5 (good)

Next, the sum of the ratings from all stations is totaled by reservoir, divided by the sum of the maximum potential ratings for each reservoir, and expressed as a percentage. This yields a possible overall health range from 20 percent (all indicators rated poor) to 100 percent (all indicators rated good) for each reservoir, regardless of the number of stations.

The scoring range then is divided into categories of good, fair, and poor ecological health, as follows:

• Ratings for run-of-river and tributary reservoirs are examined separately for apparent groupings.

• Reservoirs falling near the boundaries of the groupings are examined to establish initial break points for good, fair, and poor overall health ratings, drawing on knowledge of the historical conditions in those reservoirs.

• Those break points are then compared with a trisection of the possible overall scoring range (i.e., good–fair break point at 72 percent, fair–poor at 54 percent). Reservoirs falling near the boundaries established in both manners are examined to help adjust the break points up or down slightly to align with professional judgment.

This procedure yielded the following scoring ranges for reservoir ratings for 1993 data:

	Poor	Fair	Good
Run-of-the-river reservoirs	<52 percent	52–72 percent	>72 percent
Tributary, storage reservoirs	<57 percent	57–72 percent	>72 percent

These ranges are very similar to those developed in the same manner for the 1991 and 1992 results. The difference in break points for the poor-to-fair scoring ranges between the two types of reservoirs is because two storage reservoirs with known poor conditions rated slightly higher than the break point for poor rating on run-of-the-river reservoirs. Hence, that break point for tributary storage reservoirs was shifted upward to 57 percent.

Stream Scoring Criteria

A similar methodology is used to assess the overall ecological health at each stream monitoring location. Particular emphasis is given to the relationship between conditions at stream stations and their potential impact on downstream reservoirs.

The evaluations consider four indicators: total phosphorus, a measure of nutrient enrichment and potential for excessive algal productivity; sediment quality; benthic community; and fish community. At each station, each indicator is rated poor, fair, or good, with a numeric value of 1, 3, or 5, respectively. Equal

TABLE 8   Phosphorus Concentration and Corresponding Nutrient Ratings

Average Total Phosphorus Concentration[a]	Nutrient Enrichment Rating
< 0.05 mg/l	5 (good)
0.05 to 0.10 mg/l	3 (fair)
> 0.10 mg/l	1 (poor)

[a]In addition, waters that receive high nitrogen concentrations in the presence of sufficient phosphorus often stimulate the growth of algae and other aquatic plants to an undesirable extent. Average nitrate plus nitrite nitrogen concentrations greater than 0.65 mg/l are high relative to most Tennessee Valley streams and result in lowering a rating from good to fair or from fair to poor.

weights are given to each indicator. Scores are summed to produce an overall stream health rating ranging from 4 to 20. A station with an overall rating of 9 or less (≤45 percent) was rated poor; 10–15 (50–75 percent) fair; and 16–20 (80–100 percent) good.

Beginning in 1994, the sediment-quality rating was dropped from stream health evaluations. The sampling locations are areas where sediments accumulate only for very short periods, and evaluation of a 7-year data set shows that results do not significantly affect overall ratings.

Stream Nutrients

Phosphorus is most often the essential nutrient least available to plants relative to their needs in freshwater ecosystems; thus, low levels can limit algal productivity. When present in sufficient amounts and combined with sufficient nitrogen, phosphates may stimulate algae and other aquatic plant growth to undesirable levels. To prevent these conditions from developing in lakes, EPA recommends that total phosphorus concentrations not exceed 0.10 mg/l for streams or flowing waters or 0.05 mg/l at the point where streams enter lakes or reservoirs (United States Environmental Protection Agency, 1986). These guidelines are the basis for the ratings of stream nutrient enrichment (Table 8).

Stream Sediment Quality

The same methodology is used to evaluate stream sediment quality and reservoir sediment quality. The scoring criterion is based on a rating for the acute toxicity of sediment pore water to both the rotifer *Brachionus calyciflorus* and the daphnid *Ceriodaphnia dubia*, and a rating for the sediment concentrations of heavy metals, PCBs, organochlorine pesticides, and un-ionized ammonia. The final sediment quality rating is the average of these two ratings.

BOX 2   Benthic Community Metrics

Taxa Richness and Community Composition
- Taxa richness
- Occurrence of intolerant snail and mussel species[a]
- Number of mayfly (Ephemeroptera) taxa
- Number of stonefly (Plecoptera) taxa
- Number of caddisfly (Trichoptera) taxa
- Total number of EPT taxa[a]
- Percent oligochaetes
- Percent in the two most dominant taxa

Trophic and Functional-Feeding Group
- Percent omnivores and scavengers
- Percent collector-filterers
- Percent predators

Abundance
- Total abundance of individuals (combined quantitative samples, lower score given for extremely low or extremely high values)

[a]Metric applied to qualitative and quantitative samples combined. All other metrics applied to individual quantitative samples and resultant scores averaged.

Benthic Community

A modified version of the Benthic Index of Biotic Integrity (Kerans et al., 1992) is used to rate the condition of stream benthic communities. Twelve benthic community attributes in three areas (Box 2) are scored based on expected conditions at reference sites with water quality supportive of healthy benthic communities. Sampling effort consists of three Surber (riffle), three Hess (pool), and one qualitative sample. Metrics for EPT and intolerant snail and mussel species are computed by pooling all qualitative and quantitative samples. Total abundance is computed by pooling all quantitative samples. Metrics are computed separately for each quantitative sample at a station, then averaged by station.

The value obtained for each metric is scored poor, fair, or good (1, 3, or 5) against the best expected values for the reference sites. Scores are then summed to produce an index for each station that ranges from 12 to 60. That overall benthic community index for each stream location is classified as poor (score <30), fair (score 34–44), or good (score >45). Professional judgment is used to rate index scores between 30 and 33 as poor or fair.

BOX 3   Index of Biotic Integrity Metrics

Stream Fish Species Richness and Composition
* Number of native species
* Number of darter species
* Number of native sunfish species (excluding *Micropterus* sp.)
* Number of sucker species
* Number of intolerant species
* Percentage individuals as tolerant species

Stream Fish Trophic Structure
* Percentage omnivores
* Percentage specialized insectivorous minnows and darters
* Percentage piscivores

Stream Fish Abundance and Condition
* Catch rate (average number per unit of sampling effort, seine hauls, and electrofishing)
* Percentage individuals as hybrids
* Percentage individuals with poor condition, injury, deformity, disease, or other anomaly

---

Stream Fish Community

A modified version of Karr's (1981) Index of Biotic Integrity is used to assess the condition of the resident fish community at monitoring locations. An index and rating are produced for each site by applying the 12 metrics in Box 3.

Actual values obtained for each of these metrics are scored poor, fair, or good (1, 3, or 5) against values that would be expected at reference sites. The 12 metric scores are summed to produce an index ranging from 12 to 60, and the fish community at the stream sampling location is rated as poor (index <36), fair (index 40–44), or good (index >46). Professional judgment is used to rate fish community index values between 36 and 40 and index values of 45. These determinations are based on factors such as which of the 12 metrics rates poorest, the condition of the coexisting macroinvertebrate community, and previous Index of Biotic Integrity ratings for the station.

RESULTS

The overall ecological health ratings for the 11 run-of-river reservoirs ranged from 58 to 88 percent in 1993 (Figure 5). Of the 11 reservoirs, 4 rated good (75–

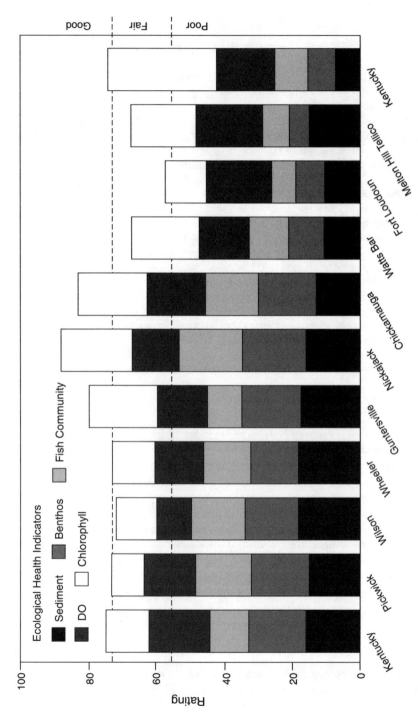

FIGURE 5 Overall ecological health of run-of-the-river reservoirs in the Tennessee Valley in 1993.

88 percent), 3 rated fair to good (71–73 percent), 3 rated fair (63–68 percent), and 1 rated poor to fair (58 percent). Overall ratings for the 19 tributary reservoirs ranged from 52 to 72 percent (Figure 6). Two reservoirs rated fair to good (both 72 percent), 14 rated fair (58–67 percent), and 3 rated poor to fair (52–56 percent).

Stream monitoring results showed a wide range of ecological conditions among the 12 streams. Three, the Clinch, Powell, and Little Tennessee Rivers, had the highest possible scores for all four ecological health indicators (nutrients, sediment, benthic macroinvertebrates, and fish community). The lowest score (50 percent) was for the French Broad River, where nutrients and fish rated poor, benthos rated fair, and sediments rated good. Scores for the remaining eight streams were evenly distributed within this range.

Most streams and reservoirs had ratings comparable to those observed in 1991 and 1992. Tributary reservoirs had generally poorer ratings than run-of-the-river reservoirs, primarily because of low DO in the hypolimnion. This is an ecologically undesirable condition that is partly due to the strong thermal stratification that occurs in deep reservoirs with relatively long retention times.

SUMMARY

This approach to stream and reservoir monitoring has proved to be a very effective way of tracking water resource conditions throughout the Tennessee Valley. The evaluation procedure focuses on critical indicators of environmental conditions and summarizes results in easily understandable terms. The information that the program produces effectively communicates information to the public and decision makers, and the technical basis for collecting and analyzing data is readily available for those who require more detailed information. TVA distributes about 200 copies of a technical summary report each year in response to requests from other agencies and individuals. The nontechnical report targeted to lake users, property owners, and the general public is mailed to about 12,000 people who have requested it, and another 40,000 copies are distributed through marinas and other public-use areas. The public's response to these products has been overwhelmingly positive.

Each year, TVA critically reviews the results to ensure that the monitoring stations are properly located and that the evaluation scheme yields useful information. As a result of these reviews, several stations have been moved slightly to be more representative of the areas of the reservoirs being sampled, and several refinements have been made of the ecological health-rating criteria. Additional refinements are expected as more information is collected.

One area not addressed by this monitoring and evaluation scheme is the level of satisfaction attained by the people who use TVA lakes and the streams that feed them. Factors such as fishing success, shoreline vistas, ease of access, and degree of solitude experienced greatly affect the aesthetic quality of recreational

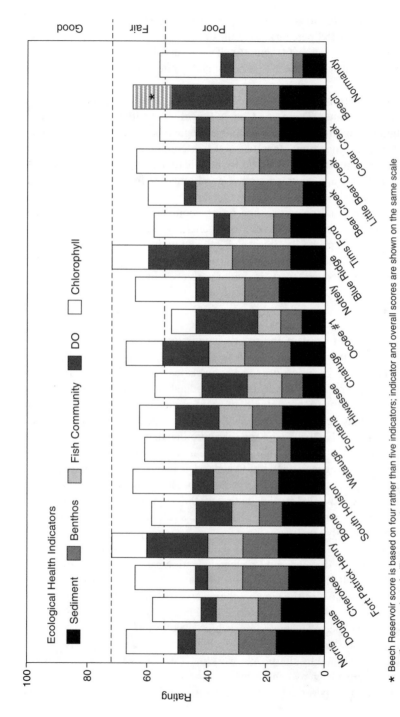

FIGURE 6 Overall ecological health of tributary reservoirs in the Tennessee Valley in 1993.

* Beech Reservoir score is based on four rather than five indicators; indicator and overall scores are shown on the same scale as other reservoirs to facilitate comparisons.

visits. The public is keenly interested in this type of information, and TVA is investigating ways to acquire it and make it available in a form similar to the evaluations of ecological health, fish consumption, and bacteriological quality.

REFERENCES

Brown, A. M., G. D. Jenkins, and G. D. Hickman. 1993. Reservoir Monitoring—1992: Summary of Fish Community Results. Norris, Tenn.: Tennessee Valley Authority.

Burns, E. R., A. L. Bates, and D. R. Webb. 1992. Aquatic Plant Management Status and Seasonal Workplan—1992. Muscle Shoals, Ala.: Tennessee Valley Authority.

Burns, E. R., A. L. Bates, and D. R. Webb. 1993. Aquatic Plant Management Program—Status and Seasonal Workplan, 1993. Muscle Shoals, Ala.: Tennessee Valley Authority.

Burns, E. R., A. L. Bates, and D. R. Webb. 1994. Aquatic Plant Management Program—Current Status and Seasonal Workplan, 1994. Muscle Shoals, Ala.: Tennessee Valley Authority.

Carlson, R. E. 1977. A trophic state index for lakes. Limnology and Oceanography 22:361–369.

Dycus, D. L., and D. L. Meinert. 1991. Reservoir Monitoring—1990: Summary of Vital Signs and Use Impairment Monitoring on Tennessee Valley Reservoirs. TVA/WR-91/1. Chattanooga, Tenn.: Tennessee Valley Authority.

Dycus, D. L., and D. L. Meinert. 1992. Reservoir Vital Signs Monitoring—1991: Summary of Vital Signs and Use Impairment Monitoring on Tennessee Valley Reservoirs. TVA/WR-92/8. Chattanooga, Tenn.: Tennessee Valley Authority.

Dycus, D. L., and D. L. Meinert. 1993a. Reservoir Monitoring—Monitoring and Evaluation of Aquatic Resources Health and Use Suitability in Tennessee Valley Authority Reservoirs. TV-93/15. Chattanooga, Tenn.: Tennessee Valley Authority.

Dycus, D. L., and D. L. Meinert. 1993b. Reservoir Monitoring—1992: Summary of Vital Signs and Use Suitability Monitoring on Tennessee Valley Reservoirs. Chattanooga, Tenn.: Tennessee Valley Authority.

Dycus D. L., and D. L. Meinert. 1994. Tennessee Valley Reservoir and Stream Quality—1993: Summary of Vital Signs and Use Suitability Monitoring, Vol. 1. Chattanooga, Tenn.: Tennessee Valley Authority.

Hickman, G. D., A.M. Brown, and G. Peck. 1994. Tennessee Valley Reservoir and Stream Quality—1993: Summary of Reservoir Fish Assemblage Results. Norris, Tenn.: Tennessee Valley Authority.

Hutchinson, G. E. 1957. A Treatise on Limnology: Vol. 1—Geography, Physics, and Chemsitry. New York: John Wiley & Sons.

Karr, J. R. 1981. Assessment of biotic integrity using fish communities. Fisheries 6(6):21–27.

Karr, J. R., K. D. Fausch, P. L. Angermier, P. R. Yant, and I. J. Schlosser. 1986. Assessment of Biological Integrity in Running Water: A Method and Its Rationale. Illinois Natural History Survey Special Publication, no. 5. Champaign, Ill: Illinois Natural History Survey.

Kerans, B. L., J. R. Karr, and S. A. Ahlstedt. 1992. Aquatic invertebrate assemblages: Spatial and temporal differences among sampling protocols. Journal of the North American Benthological Society 11(4):377–390.

Masters, A., and T.A. McDonough. April 1993. Tennessee Valley Authority, Chattanooga, Tenn. Personal communication.

Meinert, D. L., S. R. Butkus, and T. A. McDonough. 1992. Chickamauga Reservoir Embayment Study—1990. TVA/WR-92/28. Chattanooga, Tenn.: Tennessee Valley Authority.

SAS Institute. 1989. SAS/STAT User's Guide, Version 6, 4th ed., Vol. 1. Cary, N.C.: SAS Institute.

Scott, E. M., G. D. Hickman, and A. M. Brown. 1992. Reservoir Vital Signs Monitoring—1991: Fish Community Results. TVA-92/5. Norris, Tenn.: Tennessee Valley Authority.

U.S. Environmental Protection Agency. 1977. Guidelines for the Pollutional Classification of Great Lakes Harbor Sediments. Chicago, Ill.: U.S. Environmental Protection Agency.

U.S. Environmental Protection Agency. 1986. Quality Criteria for Water: 1986. EPA-44015-86-001. Washington, D.C.: U.S. Environmental Protection Agency.

Biographical Data

BRADEN R. ALLENBY is vice president for environment, health, and safety at AT&T. He joined AT&T in 1983 as a telecommunications regulatory attorney and was an environmental attorney for AT&T from 1984 to 1993. During 1992, he was the J. Herbert Hollomon Fellow at the National Academy of Engineering in Washington, D.C. Allenby is currently the vice-chair of the Institute for Electrical and Electronics Engineers Committee on the Environment; a member of the U.S. Department of Energy (DOE) Task Force on Alternative Futures for the DOE national laboratories; a member of the National Research Council Committee on Research and Peer Review in the Environmental Protection Agency; a member of the Advisory Committee on the United Nations Environmental Programme Working Group on Product Design for Sustainability; a member of the editorial boards of *The Journal of Industrial Ecology* and *Total Quality Environmental Management*; and a former member of the Secretary of Energy's Advisory Board. He is coeditor of *The Greening of Industrial Ecosystems* (National Academy Press, 1994) and is coauthor of two engineering texts, *Industrial Ecology* (Prentice-Hall, 1995) and *Design for Environment* (Prentice-Hall, 1996). Allenby teaches industrial ecology at the Yale University School of Forestry and Environmental Studies, design for environment at the University of Wisconsin, and has lectured at a number of universities, including Dartmouth College, Harvard, Massachusetts Institute of Technology, Princeton, Rutgers, and Tufts. He is a fellow of the Royal Society for the Arts, Manufactures and Commerce. Allenby holds a J.D. from the University of Virginia Law School and a Ph.D. in environmental sciences from Rutgers University.

JESSE H. AUSUBEL is director of the Program for the Human Environment and senior research associate at The Rockefeller University, as well as a program officer of the Alfred P. Sloan Foundation. His interests include environmental science and technology and industrial evolution. From 1989 to 1993, Ausubel served concurrently at The Rockefeller University and as director of studies for the Carnegie Commission on Science, Technology, and Government. From 1983 to 1988, he served as director of programs for the National Academy of Engineering. Prior to that, Ausubel served as a staff officer with the National Research Council Board on Atmospheric Sciences and Climate. He was one of the principal organizers of the first U.N. World Climate Conference held in Geneva in 1979. From 1979 to 1981, Ausubel led the Climate Task Force of the International Institute for Applied Systems Analysis, an East-West think-tank in Laxenburg, Austria, created by the U.S. and Soviet academies of sciences.

LESLIE W. AYRES worked as a systems and computer scientist at the Massachusetts Institute of Technology, IBM, and UNIVAC from 1953 through 1969. In 1970, she joined a small consulting firm in Washington, D.C., as an applications programmer, mainly on energy and environmental analysis. From 1977 to 1986, Ayres was president of Variflex Corp. and worked on various other projects, mainly in association with Robert U. Ayres. From 1987 to 1990, she was employed in the computer services department at the International Institute for Applied Systems Analysis in Laxenburg, Austria. Since 1992, Ayres has been a research associate in the Center for Management of Environmental Resources at the European Business School INSEAD, Fontainbluau, France, where she has coauthored several papers and a book, *Industrial Ecology: Towards Closing the Materials Cycle* (Edward Elgar Publishers, 1996).

ROBERT U. AYRES is Sandoz Professor of Environment and Management, professor of economics, and director of the Centre for the Management of Environmental Resources at the European Business School INSEAD in Fontainebleau, France. From 1979 to 1992, he was professor of engineering and public policy at Carnegie Mellon University in Pittsburgh. Ayres holds a Ph.D. in mathematical physics from Kings College, University of London. He has been affiliated with the Hudson Institute, Resources for the Future, and the International Institute for Applied Systems Analysis. Ayres has published more than 150 journal articles and book chapters and has authored or coauthored 12 books on topics ranging from technological change, manufacturing, and productivity, to environmental and resource economics. His most recent books are *Industrial Metabolism: Restructuring for Sustainable Development* (UNU Press, 1994), *Information, Entropy, and Progress: A New Evolutionary Paradigm* (AIP Press, 1994), and *Industrial Ecology: Towards Closing the Materials Cycle* (Edward Elgar Publishers, 1996).

NEIL E. CARRIKER is the quality manager for the Tennessee Valley Authority (TVA) Water Management Division. Previously, he was a senior environmental engineer in the same organization with responsibilities for a variety of activities focusing on monitoring the quality of water in the Tennessee River and its reservoirs and tributaries, and interpreting the data to identify trends and relate water quality to land use, natural processes, and pollution impacts. Carriker represents TVA on several interagency committees and task groups.

CRAIG COX is special assistant to the chief of the Natural Resource Conservation Service of the U.S. Department of Agriculture (USDA), where he is responsible for strategic planning and natural resource assessment. He was formerly a senior staff officer at the Board on Agriculture of the National Research Council (NRC), where he directed three major studies, including Soil and Water Quality: An Agenda for Agriculture, and Rangeland Health: New Methods of Classifying, Inventorying, and Monitoring Rangelands. Between his stints at the NRC and USDA, Cox was on the staff of the Senate Committee on Agriculture, Nutrition and Forestry, where he led work on natural resource and environmental issues. He holds an M.S. in agricultural and applied economics from the University of Minnesota.

JOHN R. EHRENFELD is senior research associate in the Massachusetts Institute of Technology (MIT) Center for Technology, Policy, and Industrial Development. At MIT since 1985, he directs the Program on Technology, Business, and Environment. Ehrenfeld also serves as a core faculty member in the MIT Technology and Policy Program. His research examines the way businesses manage environmental concerns; systems for introducing design for environment into the product development process; the impacts of voluntary codes of corporate environment management on strategy development and culture change; and industrial ecology. Ehrenfeld is a member of the American Chemical Society, American Association for the Advancement of Science, Air & Waste Management Association, and the Society for Risk Analysis. He is an editor of the new *Journal of Industrial Ecology* and a member of the editorial advisory board of *Environmental Science & Technology*. Ehrenfeld holds a B.S. and Sc.D. in chemical engineering from MIT and is author or coauthor of over 100 papers, reports, and other publications.

PAUL FAETH is a senior associate in the World Resources Institute (WRI) Program in Population and Economics. He directs a project area on the economics of sustainable agriculture. The first phase of this research involved a multi-country effort examining the impact of agricultural policies on the adoption and generation of resource-conserving agricultural technologies. The second phase was a national natural resource accounting study of U.S. agriculture—the first ever of its kind. Faeth is WRI's liaison to the Sustainable Agriculture Task Force of the President's Council on Sustainable Development. He directed WRI's effort to help a power company mitigate its CO_2 emissions through forestry activities in developing countries. This effort resulted in the first project ever to be funded with the intention of balancing such emissions. He worked previously with the International Institute for Environment and Development, where he applied methods of systems analysis to examine the environmental impacts of development projects. Faeth has also worked for the U.S. Department of Agriculture's Economic Research Service on issues related to agricultural trade policy. He is a member of the American Society of Agricultural Engineering, the American Agricultural Economics Association, and the American Economic Association. Faeth holds degrees in agricultural engineering from the University of Florida and in resource policy from Dartmouth College.

FRANK R. FIELD III is senior research engineer at the Massachusetts Institute of Technology's (MIT's) Center for Technology, Policy, and Industrial Development. He is also director of the Materials Systems Laboratory, a leading research group in the development of analytical methods in materials systems analysis, which is the application of engineering and economic principles to problems in materials use, substitution, and processing. Field's research focuses on the development and application of decision analysis tools to problems in materials selection and substitution. He has examined problems of materials competitiveness in a wide range of engineering areas, including automotive, aerospace, and electronic applications. Field, who holds a Ph.D. from MIT, also teaches strategic planning and materials systems analysis and materials policy. He was appointed as lecturer in technology and policy at MIT in 1994.

ROBERT A. FROSCH is a senior research fellow at the John F. Kennedy School of Government, Harvard University. He recently retired as vice president of the General Motors Research Laboratories. Frosch's career combines varied research and administrative experience in industry and government service. He has been involved in global environmental research and policy issues at both the national and the international level. From 1951 to 1963, Frosch was employed at the Hudson Laboratories of Columbia University, first as a research scientist and then as director from 1956 to 1963. In 1963, he became director for nuclear test detection in the Advanced Research Projects Agency (ARPA) of the U.S. Department of Defense and in 1965 became ARPA deputy director. In 1966, Frosch was appointed assistant secretary of the Navy for research and development. He served in that position until January 1973, when he became assistant executive director of the United Nations Environment Programme. In 1975, Frosch assumed the post of associate director for applied oceanography at the Woods Hole Oceanographic Institution. From 1977 to 1981, he served as administrator of the National Aeronautics and Space Administration. Frosch served as president of the American Association of Engineering Societies from 1981 to 1982 and is a member of the National Academy of Engineering. He holds a Ph.D. in theoretical physics from Columbia University.

CLYDE E. GOULDEN is a full curator of the Academy of Natural Sciences of Philadelphia and an adjunct professor of biology at the University of Pennsylvania. His research focuses on zooplankton nutrition and ecotoxicology. Recently, Goulden's nutritional research has emphasized the importance of dietary lipids. During the past 15 years, he has used several different bioassay protocols to study the toxicity of effluents to *Daphnia*, and he has participated in several Environmental Protection Agency workshops on developing aquatic bioassay protocols.

THOMAS E. GRAEDEL is a distinguished member of the technical staff at Bell Laboratories, Lucent Technologies (formerly AT&T), where he has been employed for 27 years. He serves currently as co-chair of Lucent's Design for Environment Assessment Team. Graedel's areas of specialty, in which he has more than 200 publications, including 7 books, are atmospheric corrosion, atmospheric chemistry, and environmentally conscious manufacturing. He was the first atmospheric chemist to perform computer-model studies of the gas-phase chemistry of sulfur and the interactive chemistry of raindrops, and to study the reactions involved in atmospheric corrosion. In connection with this latter specialty, Graedel served as corrosion consultant to the Statue of Liberty Restoration Project from 1984 to 1986. Graedel is the coauthor of the textbook *Atmospheric Change: An Earth System Perspective* (W.H. Freeman, 1993), which is being used at more than 25 universities around the world. His latest books, *Industrial Ecology* (Prentice Hall, 1995) and *Design for Environment* (Prentice Hall, 1996), coauthored by B. R. Allenby, are the first engineering-design texts to consider environmental impacts over the entire product and process life cycle.

MARTIN B. HOCKING is an associate professor in the Department of Chemistry at the University of Victoria in British Columbia. He has teaching and research interests in the areas of industrial and environmental chemistry, organic chemistry, and polymer synthesis, and has published numerous papers related to these interests. Hocking holds nine patents in the fields of monomers, process chemistry, and medical devices. He chaired the senate committee that initiated the Environmental Studies Program at the University of Victoria and headed the new program for a year. Hocking's experience in industry prior to his appointment at the University of Victoria was the source of some of the material assembled in his book, *Modern Chemical Technology and Emission Control* (Springer-Verlag, 1985). He was associate editor and a contributor to the volume *Effects of Mercury in the Canadian Environment*, a publication of the National Research Council of Canada. Hocking has also conducted research at the Pulp and Paper Research Institute of Canada, McGill University, University College

London, and the University of New South Wales, Sydney, Australia. He holds a
Ph.D. from the University of Southampton.

SUSAN E. OFFUTT is administrator of the U.S. Department of Agriculture
Economic Research Service (ERS). Prior to heading ERS, Offutt was the executive director of the Board on Agriculture and assistant executive officer at the
National Research Council. Before taking over at the board in 1992, she was
chief of the agriculture branch at the Office of Management and Budget. Offutt
has taught econometrics and public policy in the agricultural economics department at the University of Illinois. Her research interests include commodity market instability and structure and the economics of the development and adoption
of new technologies. Offutt holds a B.S. from Allegheny College and an M.S.
and Ph.D. from Cornell University.

EDWARD T. RANKIN is a stream ecologist with the Ecological Assessment
Unit of the Ohio Environmental Protection Agency. His responsibilities include
developing biological criteria for streams, developing habitat assessment methods, and compiling the Ohio Water Resource Inventory. Rankin holds an M.S. in
zoology from Ohio State University.

ROBERT C. REPETTO is vice president of the World Resources Institute
(WRI) and directs its Program in Economics and Population. He is the author of
numerous publications on the environment and economics. Repetto is a member
of the Environmental Protection Agency's Science Advisory Board and the National Research Council's Board on Sustainable Development. Before joining
WRI in 1983, he was an associate professor in the School of Public Health at
Harvard University. Previously, Repetto was a resident advisor for the World
Bank Mission in Indonesia, economic advisor to the planning and development
board for the government of East Pakistan, staff economist for the Ford Foundation in New Delhi, and an economist for the Federal Reserve Bank of New York.
He holds a Ph.D. in economics from Harvard University.

LOUIS SAGE was formerly the vice president of the Academy of Natural
Sciences in Philadelphia and director of its Division of Environmental Research.
In those positions, he had responsibility for the Stroud Center, Patrick Center,
Benedict Center, and the Maritza Center in Costa Rica. Sage has served in numerous capacities on the Chesapeake Bay Program, including as a member of the
Scientific and Technical Advisory Committee for 11 years. His present activities
include service on the boards of various organizations, among them the Institute
for Cooperation in Environmental Management, the National Water Alliance,
and the Alliance for the Chesapeake Bay, where he is currently serving as
president.

PETER C. SCHULZE is assistant professor of biology, Austin College, Sherman, Texas. He teaches courses in ecology and environmental studies. Before joining the faculty of Austin College, Schulze held a postdoctoral appointment at Lehigh University and taught at Dartmouth College and Harvard University. He was the 1993–1994 J. Herbert Hollomon Fellow of the National Academy of Engineering. Schulze's primary interests are in aquatic ecology and impacts on the environment. He edited the volume *Engineering within Ecological Constraints* (National Academy Press, 1996). Schulze holds a Ph.D. in biology from Dartmouth College and has been recognized by Harvard University for distinction in teaching.

ARTHUR J. STEWART is a senior research staff member in the Environmental Sciences Division at the Oak Ridge National Laboratory and an adjunct faculty member of the University of Tennessee's Department of Ecology and Evolutionary Biology. He was previously an Oak Ridge Associated Universities postdoctoral fellow and an assistant professor in the Department of Botany and Microbiology at the University of Oklahoma. Stewart's research focuses on ecotoxicology, biological monitoring, bioassay development, and stream ecology. He has authored or coauthored more than 50 journal articles and numerous technical manuscripts and reports. Stewart is on the editorial boards of *Environmental Toxicology and Chemistry* and *Ecotoxicology*. He has won numerous prizes for his poetry and is presently editing his first book of poems, which will be published under the title *Random Holes in Darkness*. Stewart holds a Ph.D. in limnology from Michigan State University.

RICHARD STRANG is a technical associate with Eastman Chemical. His work with the company's Health, Safety and Environmental Services organization includes environmental studies for Eastman facilities in Arkansas, South Carolina, Tennessee, and Texas. Strang represents Eastman in the Water Environment Federation, The Society for Risk Analysis, and on environmental risk committees for the American Industrial Health Council and the Chemical Manufacturer's Association.

REBECCA TODD is associate professor of accounting in the Boston University Graduate School of Management. Previously, she taught in graduate programs at the University of North Carolina at Chapel Hill and the Stern School at New York University. Todd has developed MBA courses in financial statement analysis as well as executive programs in accounting and financial analysis, analysis of derivatives disclosures, international financial statement analysis, analysis of financial institutions, and other topics. She is especially interested in the use of environmental accounting and financial analysis to track, analyze, and report on environmental costs in order to better manage wastes and other substances. A number of Todd's case studies and analyses have been published in journals and handbooks. She received her CFA (Chartered Financial Analyst) charter in 1990 and since that time has taught Level I and Level II financial statement analysis in CFA review programs in the United States, Zurich, and the Far East. Todd is chief consultant for the development of the Association of Investment Management's Equity Specialization Program, a post-CFA curriculum. She holds a Ph.D. in business administration from the University of North Carolina at Chapel Hill.

IDDO K. WERNICK is associate research scientist at the Earth Institute, Columbia University. His current research covers long-term patterns of natural resource use in the United States and the resulting environmental effects. This work concentrates on analyzing the flows of materials in the U.S. economy. He has also investigated environment-related causes of mortality and the technical and political context for community risk assessment. Wernick was formerly a research associate with the Program for the Human Environment at The Rockefeller University. He holds a Ph.D. in applied physics from Columbia University.

JOHN WESTRA is a graduate research assistant in the Department of Applied Economics at the University of Minnesota, where he is studying the adoption of conservation tillage practices by farmers along the Minnesota River. From 1992 to 1994, he worked as a research analyst in the Economics and Population Program of the World Resources Institute (WRI). In that capacity, Westra helped develop an economic model that incorporated the on-site and off-site environmental impacts of conventional and sustainable production practices. Before joining WRI, he worked as an agricultural economist with the U.S. Department of Agriculture and as a freshwater fisheries technician for the Peace Corps in Guatemala. Westra holds an M.S. in agricultural and resource economics from the University of Maine.

CHRIS O. YODER is environmental manager, ecological assessment, for the Ohio Environmental Protection Agency's Division of Surface Water. He is responsible for the ecological evaluation of Ohio's streams, rivers, and wetlands. Yoder has developed biological, chemical, and physical assessment methods and criteria for these habitats and Lake Erie. He holds an M.A. in zoology.

THOMAS W. ZOSEL is manager, environmental initiatives, for 3M Corporate Environmental Technology and Services. He has been with 3M for 29 years, 23 of which have been in the environmental area. Zosel is responsible for following major environmental legislative and regulatory activity, communicating its impact on 3M, and developing a proactive response to assure beyond-compliance performance. He also interfaces with 3M's marketing community to develop customer-related initiatives and coordinate 3M's environmental stakeholder communications. Zosel currently serves on the U.S. Environmental Protection Agency's Clean Air Act Advisory Committee, is past chair of the American Institute of Chemical Engineers Center for Waste Reduction Technologies, chairs the National Pollution Prevention Center Advisory Board, and is a frequent author and speaker on pollution prevention and industry's proactive response to environmental issues.

Index